Global Climate Change and California

GLOBAL CLIMATE CHANGE AND CALIFORNIA

POTENTIAL IMPACTS AND RESPONSES

Edited by Joseph B. Knox
with Ann Foley Scheuring

UNIVERSITY OF CALIFORNIA PRESS
Berkeley   Los Angeles   Oxford

The publisher wishes to acknowledge the generous assistance of the National Institute for Global Environmental Change in the publication of this book.

University of California Press
Berkeley and Los Angeles, California

University of California Press
Oxford, England

Library of Congress Cataloging-in-Publication Data

Global climate change and California : potential impacts and responses / edited by Joseph
    B. Knox with Ann Foley Scheuring.
        p.   cm.
    Includes bibliographical references.
    ISBN 0-520-07555-2 (cloth : alk. paper). — ISBN 0-520-07660-5 (paper : alk. paper)
        1. Climatic changes—California.   2. Greenhouse effect, Atmospheric—California.
    3. California—Climate.   I. Knox, Joseph B.   II. Scheuring, Ann Foley.
    QC981.8.c65g57   1991
    551.69794—dc20   91-24538
                                                                                  CIP

Printed in the United States of America

1 2 3 4 5 6 7 8 9

The paper used in this publication meets the minimum requirements of American
National Standard for Information Sciences—Permanence of Paper for Printed Library
Materials, ANSI z39.48-1984 ∞

CONTENTS

ACKNOWLEDGMENTS

In offering this volume for public distribution, the editors wish to acknowledge the financial support of the U.S. Department of Energy—
Office of Environmental Policy Analysis, the Office of the President of
the University of California, the Save the Earth Foundation, and the
Universitywide Energy Research Group, as well as the organizational
skills of Noreen Dowling and the staff of the Public Service Research
and Dissemination Program at the University of California, Davis. Marcia
Kreith made the workshop arrangements and prepared materials for
distribution. Kelly Carner proofed and corrected drafts and checked
final copy. Shannon Fox typed and formatted the document.

Each of the working groups at the initial conference developed working papers, to which several dozen conference participants contributed
in the critiquing process. The leaders of these working groups* are to
be thanked for their leadership and success in encouraging quality participation. The Energy Caucus leader, Allen Edwards, and the Climate
Caucus leader, Michael C. MacCracken, identified common research
needs across all the working groups. The many reviewers of the various
chapter drafts, and the product itself, represent the collected expertise
of researchers in many disciplines. We are indeed indebted to them all.

*Lowell N. Lewis and D. William Rains (Agriculture); Henry J. Vaux, Jr. (Water Re-
sources); Daniel B. Botkin (Natural Ecosystems); Richard A. Berk (Human Dimensions).

ONE

Global Climate Change:
Impacts on California
An Introduction and Overview

Joseph B. Knox

In the fall of 1988 the University of California organized a new public-service initiative on global climate change in response to inquiries and requests from members of Congress and the Department of Energy (DOE). This new systemwide initiative involved all of the University of California campuses and the University's three national laboratories at Berkeley, Los Alamos, and Livermore. The goal of this Greenhouse Initiative was to focus the multidisciplinary resources of the UC campuses and the team-oriented research capabilities of the laboratories on the prospect of global warming and its associated effects on the planet and its nations. This issue may well be the most challenging environmental issue facing the world in the 1990s and beyond.

The Chancellor's Office in conjunction with the Public Service Research and Dissemination Program at the University of California, Davis, served as the coordinating agency for this new initiative. Early in the organizing process, the decision was made to begin activities through a stepwise workshop process. In consultation with the DOE, the organizers proposed a series of workshops to focus University of California research resources on the issue of global warming, to contribute to the congressionally mandated DOE studies on options for the U.S. to reduce carbon dioxide emissions by 20 percent by the year 2000, and to begin building a long-term research base contributing to an improved understanding of global change in all of its complexity and diverse discipline implications.

AN OVERVIEW OF THE WORKSHOP PROCESS

In consultation with many people, we organized three workshops during 1989 to assure that findings would be available in a timely fashion for the national dialogue on global warming and the national studies in progress.

Workshop #1: Global climate change and its effects on California.   The first workshop (July 10–12, 1989) considered the effects of a projected global warming on salient elements of California's economy, including its water resources, energy supply and demand, agriculture, forests, terrestrial and aquatic ecosystems, coastal zone, and urban areas.

Workshop #2: Energy policies to address global climate change.   The second workshop (September 6–8, 1989) focused on possible policy actions that might be technically, economically, and institutionally helpful in reducing emissions of CO_2 and other so-called greenhouse gases in the twenty-first century. Topics included both short-term solutions involving energy efficiency (present–2010) and longer-term energy alternatives.

Workshop #3: Pacific Rim research development strategies related to climate change.   The third workshop (October 22–27, 1989) was designed to contribute to the building of the new level of international understanding and cooperation required to address the challenge of global change. This event assembled representatives from Pacific Rim nations, including those who own most of the world's coal, those whose tropical forests are being destroyed, those impacted by rising sea levels, and those with rapidly growing populations. Many of the Pacific Rim nations share a common goal, to achieve a better standard of living or quality of life for their citizens, even in the midst of potential climate change. Recognizing that there may be conflicting interests, including population-growth pressures, we need to build the basis for unprecedented mutual understanding and cooperation for our common future.

The chapters in this volume, *Global Climate Change and California: Potential Impacts and Responses*, are made up of papers from the first of the UC/DOE workshops. Each chapter concludes with specific recommendations for a research agenda. Development of this academic research agenda played a vital role in the preparation of the University of California for joining in a new National Institute for Global Environmental Change (NIGEC), created by Congress in September 1989, approximately one year after the initial concept of the workshop series. The sequence of events leading up to the formation of NIGEC was, however, virtually unknown to workshop participants during their endeavors. The story of the birth of NIGEC is contained in the epilogue of this volume.

Background Materials for Workshop #1

Participants in Workshop #1 were provided with a common set of background materials including a postulated, but plausible, estimate of climatic conditions for the state of California in a) the first decade of the twenty-first century, and b) for the period 2030–2070. These climate scenarios were constructed to represent plausible climate changes that are consistent with projections by climate scientists. Each of the panels in Workshop #1 were expected to base their considerations, discussions, and findings on these climate scenarios, so that the ultimate workshop findings could be intercompared.

The background materials included several overview studies of the global-warming issue, authored by the Canadian Climate Center, the well-known Toronto International Conference, and the AAAS Panel on Climate and Water, and summaries by the API (American Petroleum Institute) and others. "Greenhouse Warming: What Do We Know?" by Mike MacCracken presented a summary from the perspective of climate modelers of the evidence for global warming, our theoretical understanding of the contributing mechanisms, and discussion of the geometrical scale on which we can have confidence in the projections. Contributions by Gleick, Revelle, and Knox/Buddemeier introduced the participants to the seminal role that water resources play in the California infrastructure and how these resources might be affected by global warming. "The Changing Atmosphere—Challenges and Opportunities" contained a joint statement prepared by the UCAR (University Corporation for Atmospheric Research) Board of Trustees and the American Meteorological Society. This statement reflects the consensus of many of the leading atmospheric scientists in our country that global climate change is one of two key issues facing the atmospheric sciences. "Current Views and Developments in Energy/Climate Research" by W. Bach presented the case that energy efficiency, fuel efficiency, and conservation are plausible and necessary options if mankind wishes to pursue a future with minimum climatic risk; this argument, advanced in 1983, seems even more relevant today. We also included "greenhouse" statements by two California Congressmen, Mr. Fazio and Mr. Brown, which demonstrated the vital interest in Congress in this issue and proposed the development of national strategies in response to the threat of global warming.

The global nature of the greenhouse issue, evidenced by the projected rise in sea level for the twenty-first century, was explained in "Forecasting Changes in Sea Level Due to Greenhouse Gases." The recently completed study by NOAA (National Oceanic and Atmospheric Administration), led by Kirby Hansen, which examined surface-temperature records of U.S. climatic stations from 1985 to 1987, suggested that there has been little or no warming trend in the continental U.S. during this time. This report attracted a great deal of attention from workshop participants, particularly those who questioned the evidence of global warming based on analysis of surface-air temperature records. Advocates of the greenhouse position pointed out that the U.S. contains only 1.5 percent of the surface of the Earth, that the pattern of climate system response is complicated with the presence of some cooling areas, that much of the rest of the surface of the planet is warming, and that hence warming dominates the global trend.

Workshop Panels

Workshop #1 was divided into four panels serving as discussion and brainstorming groups. These panels were:
1) Agriculture
2) Natural Ecosystems
3) Water Resources, Soil Systems, Groundwater, and the Sacramento/ San Joaquin Delta
4) The Human Dimensions—Current Trends and Future Directions for California

Each panel chair provided an overview, a preliminary statement of issues, and some discussion directions for the panel in order to assess the effects of the postulated change in climate on California. Impacts on energy demand and natural and human systems were considered part of each panel's discussion effort. For example, if a warmer and drier summer and reduced summer runoff required more pumping of groundwater, energy demand would be increased; if summer urban temperatures were warmer or perhaps cooler along the immediate coast, the impact on energy demand should be considered; if transbasin transport of water were required, additional energy would be required.

Organizers of the workshop anticipated that most of the panels would state a strong need for higher-resolution climate-scenario information, in both space and time, than climate modelers could provide. Such requests did surface in all the panels; the resulting specific data needs were noted by the participating climate modelers, who realized that such informational needs required a coordinated approach. The "Climate Caucus" was created to provide this service to the panels, as well as to represent the needs of impact modelers to climate scientists. At least one climate modeler participated in each of the panels. This crosscutting caucus therefore served an important integrating function in development of the workshop findings.

In a similar fashion, workshop organizers created an "Energy Caucus" to provide an energy policy resource for each panel and to integrate observations on the energy implications of identified impacts on California's infrastructure. The information so gathered is reported in chapter 8 of this volume.

The goal of the workshop was to go beyond the identification of issues relevant to each panel and to identify policy actions that should be explored in the search for common ground. Our objectives were to
a) enumerate questions and needs faced by decision makers;
b) identify vulnerabilities and sensitivities in both society and the economy;

c) define the climate information required to assess impacts on the various systems;

d) explore the extent to which climate models will be able to provide needed information;

e) identify gaps in research or needed research;

f) identify the key uncertainties whose reduction would have the greatest payoff.

Projections of Future Climate Conditions

Ideally, members of the climate-modeling community would have been able to provide the panels with reliable, spatially detailed predictions of the climate variables needed to assess changes in farm output, the productivity of ecosystems, the behavior of watersheds, and air quality changes in specific airsheds in California for a time when the effective CO_2 concentration doubles (e.g., 2030–2070). For over thirty years climate modelers have been pursuing the elusive goal of representing the complex coupled ocean-atmospheric system and its cycling of radiatively active gases in ever more faithful ways. In 1989, however, we were still far from this ideal goal. Our models lack many necessary attributes, including appropriate coupling of the oceans, inclusion of detailed topography important for the Pacific coast, improved treatment of cloud processes, and adequate spatial resolution, to name a few. The modeling community is at least several years away from having an enhanced capability that includes improvements in these areas that have been tested and verified against present and past climatic conditions.

In the absence of perfect models of the temporal evolution of the climate, several methods for projecting future climate states have been proposed and investigated over the past thirty years. These include:

1) deriving estimates of future climate change by extrapolating from the instrumental record of climate fluctuations over the past century;

2) drawing analogs from climate states during historical or paleoclimatic periods;

3) interpolating from simulations by General Circulation Models (GCMs) of the equilibrium change in the climate system that would be induced by a doubling of the CO_2 concentration to about 600 ppm;

4) using coupled or nested mesoscale models within a GCM to obtain a higher spatial resolution to the simulations of the GCM;

5) coupling the use of various statistical submodels to GCMs to generate estimates of regional-scale distributions of climate variables from coarse-grid GCM outputs.

Each of these methods for generating climate change projections has its strengths and weaknesses:

1) Ingenious Use of the Instrumental Record—The limitations of this method include the fact that the land-based surface-air temperature record is flawed or impacted, in part, by the process of urbanization around the weather stations. Further, the projections of regional climate change so generated correspond to smaller changes in average surface-air temperature than those associated with the doubled atmospheric CO_2 concentration GCM simulations. The projections so generated do, however, correspond to a global warming of about $0.5°$ C, a level likely to be achieved in the time period 2000–2010. This method thus produces only a partial answer to the question of $2 \times CO_2$ climatic conditions.

2) Climatic Analogs—The principal limitation of this method is that the orbital parameter changes that induced past glacial and interglacial periods provide a significantly different forcing than changes in atmospheric composition. Confidence in such analogs as a reliable guide must therefore be limited.

3) GCM Simulations of Perturbed Climate States—The climate community readily acknowledges that the current models need extensive improvements as enumerated previously, new validations against the current climate state, and new high-resolution simulations of the time-dependent response of the climate system to the increasing atmospheric concentration of the several greenhouse gases, including CO_2. The realization of such a validated climate prediction for the "warmer world of the next century" is at least several years away.

4) Nested Mesoscale Models Coupled to GCMs—The limitations mentioned above in 3) are equally applicable to this method.

5) Statistical Submodels Coupled to GCMs—Recently there has been notable success in the statistical coupling of GCM grid-point data to water-balance models at the watershed spatial scale by Peter Gleick, for the purpose of estimating the changes in winter and summer runoffs in the Sacramento Basin for doubled CO_2 conditions. The limitations of this method include the possibility that the statistical relationships established for the current climatic state may not hold in the $2 \times CO_2$ world. Further, since the watershed models are driven by yet imperfect GCMs, or GCMs in need of marked improvement, subtle but important changes in storm tracks may not yet be included. If climatic zones are believed to shift poleward in the "warmer world," one would intuitively surmise that the major storm paths should likewise shift. Marked improvement in the projections from this method must probably await the arrival of improved, higher-resolution GCM simulations.

Based on these strengths and weaknesses, it is evident that those evaluating global change are in the unenviable position of having inadequate modeling tools, imperfect databases, and the prospect of a long period prior to the arrival of enhanced capabilities.

Scenarios for Assessing Societal Impacts

Since the mid-1980s there has developed a keen sense of urgency in the climate community that projections of the sensitivity of our energy demand, water availability, and infrastructures to global change must be a subject of serious study. We recognize that precise regional-scale climate predictions are beyond the reach of current simulation capabilities; and uncertainty necessarily exists about the spatial distribution and timing of the climatic response to greenhouse forcing as the atmospheric concentration of greenhouse gases inexorably increases. Several papers in our background material reflected this perspective. Investigations of climatic impacts have generally been based on regional climate scenarios applicable to a specified time period in which the conditions of $2 \times CO_2$ are plausible.

The word *scenario* is used above to emphasize the lack of precision in predictions of climate changes in the next century, and to reflect the considerable uncertainty in these projected climatic conditions for use in effects studies. Regional climate scenarios are nevertheless a legitimate tool for analysis and discussion. The findings of such provisional studies can be very valuable in identifying vulnerabilities, which if addressed could make our infrastructural elements more robust against natural climatic variability, population growth, and "greenhouse" impacts. Sensitivity studies can play an important role in identifying possible policy actions for the lessening of greenhouse effects in the future, or for the purpose of buying time (i.e., delaying the arrival of projected greenhouse warming). Many of the possible policy actions can be justified on grounds other than addressing the greenhouse issue; they can promote changes of value to society in any event, and hence be relatively inexpensive moderators of the greenhouse effect. It is with this rationale that our workshop process was conceived.

Infrastructure assumptions. The infrastructure assumptions utilized in the 1982 study entitled *Competition for California Water* (Engelbert and Scheuring 1982) were adapted to our workshop process. Significant studies of water resources in California from 1982 and 1984 were thus used as related references to our considerations of "Global Climate Change and Its Effects on California." University staff who participated in those important studies were able to transfer their experience to our study with a minimum of difficulty. The infrastructure assumptions are given below, with the one exception that the assumption regarding "no dramatic change in climate" is replaced by our selected climate scenarios. (Material below is a direct quotation except for 1b.)

1) No real disasters, or technical or social breakthroughs for California—for example,
   a) No world war.

b) The assumption "no dramatic climate change" is replaced by the climatic conditions of the future reflected in our proposed climate scenarios for the period 2000–2010 and for 2 × CO_2 climate in 2030–2070.

c) No unforeseen environmental threat or epidemic [this is interpreted to include no massive or extensive period of volcanism in the next seventy years (Joseph B. Knox)].

d) No collapse in political, economic, or social systems.

2) No dramatic developments significantly easing California's socio-economic problems and pressures—for example,

a) No limit on immigration from other states.

b) Continued immigration from Mexico.

c) No new cheap energy source. Rising energy prices.

d) No dramatic discovery of a new means of water supply or pollution treatment.

3) Demographic projections based upon California Department of Finance Report 77-P-3, Series E-150, 1977.

4) Continuing pressure from the danger of disruptions in international commerce, energy supply, and food supply. This includes the possibility of famine.

In spring 1989, as we prepared for our workshop, it was tempting to question some of these assumptions or add to them in some way. For example, massive oil spills can disrupt the flow of 25 percent of the U.S. domestic oil supply; easing tensions between the U.S. and USSR could lead to lower defense budgets in the two countries and hence impact California's industrial growth; bioengineering could produce revolutionary developments in the future; and the recent California droughts may have made our water-resource institutions more open to change and competing demands for water. Breakthroughs in the next seventy years will no doubt occur, but to predict them by sector is beyond the scope of our workshop.

Near-Term Scenario

Our near-term climate scenario, for the period 2000–2010 (see table 1), is constructed primarily by extrapolation from the instrumental record of surface-air temperature T_S, following the techniques of Wigley (1986) as summarized in Edmonds et al. 1986. In order to create "warm world" analogs, Wigley stratified the surface-air temperature database in four slightly different ways: in his second method, the anomaly fields for surface pressure, surface-air temperature, and precipitation were generated for each season by subtracting the seasonal-composite field for the warmest decade (1936–1945) from that of the coldest decade (1963–1972) in the data period 1925–1973. The fact that the colder decade followed the warmer decade even though greenhouse gas con-

TABLE 1.   Two Climate Scenarios Used in the Workshop

A. Near-Term Scenario: Period 2000–2010

1) Increase in average annual global temperature above 1951 to 1980 levels = 0.5° C.
2) Seasonal changes in California surface-air temperature and precipitation:

	T_s	ppt
autumn	+0.5° C	drier
winter	+0.5° to 1.0° C	wetter
spring	+0.5° C, in north	wetter
summer	location-dependent	remains dry

B. 2 × CO_2 Conditions for the Period 2030–2070

1) Increase in annual average surface temperature above 1951 to 1980 levels = 2° to 4° C.
2) Change in precipitation: +10% globally, ±20% in California.
3) Rise in mean sea level: 0.2 to 1.0 meters.
4) Rise in snow level of 100 meters for each 1° C increase in temperature.
5) Air pollution: increase in the peak surface-air concentrations of ozone downwind of urbanized areas of 10% to 20% due to the increased surface temperature, increased UV_b flux, and higher global background of tropospheric O_3.
6) The UV_b flux to the Earth's surface: 50% increase.
7) Mean temperature of shallow-water bodies: increase with average daily dry-bulb temperature.
8) Storm tracks: although very uncertain at this time, it is reasonable to assume a poleward shift in storm tracks consistent with the projected poleward shift of climatic zones.

centrations were rising is a consequence, scientists generally believe, of the larger effect of natural fluctuations, volcanism, and so on during these decades. Nonetheless, whatever the cause, the change in the mean Northern Hemispheric temperature between these two decades was 0.5° C, a level of global-scale warming that Wigley estimates could occur in the early twenty-first century. The results of these data manipulations are shown in figures 1 and 2 for surface-air temperature and precipitation. The precipitation anomaly is conveniently expressed in terms of the standard deviation and permits interpolation of the anomaly to watersheds that might indeed have quite different mean seasonal precipitation.

Figure 1 suggests that the surface-air temperature in the period 2000–2010 will be warmer in spring, fall, and winter, with the greatest increase occurring in the winter. The "warm world" construct indicates

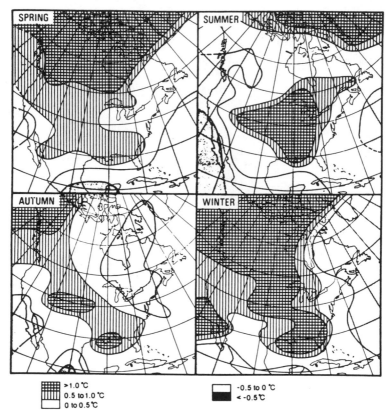

Figure 1. Temperature-change scenarios for North America. Scenario A.

that the summer season may be cooler along the coast, which at first might seem a contradiction, but could be quite plausible, if the hotter interior were to contribute to the development of a stronger sea breeze circulation through the passes of the coastal mountains.

The maps in figure 2 suggest increased precipitation in California in winter and spring by more than one-half a standard deviation, while autumn rains might be delayed. This interpretation is supported by the sea-level pressure anomaly that shows an enhanced offshore gradient in the fall season, conducive, perhaps, to more frequent Santa Ana wind conditions.

Since Wigley completed his research on the construction of climate scenarios from the instrumental record, several investigators have questioned the quality of the land-based, surface-air temperature database. The issue raised involves the extent to which urbanization has affected

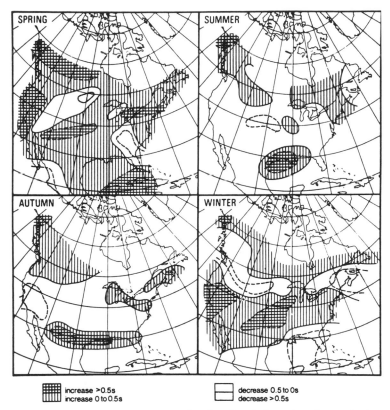

Figure 2. Precipitation-change scenarios for North America. Scenario A.

the T_S time series in the expanding populated areas (Karl and Jones 1989). In addition to urbanization, grazing practice changes, deforestation, and expansion of irrigation all have the potential to affect measurement locations through local climate modification. While urbanization is widely held to raise the mean daily temperature, expansion of irrigation could moderate both the maximum and minimum daily temperatures. Several investigators, after credible efforts to correct the T_S time series, find that the average global mean temperature has risen about 0.3° to 0.4° C during this century. We raise this point for a specific reason, in that in Wigley's construction of a climate analog (Method B), the seasonal composite of the coldest decade (1963–1972) is subtracted from the seasonal composite of the warmest decade (1936–1945). If urbanization has flawed the record, then the T_S values of the cold decade (1963–1972) are slightly high, and the difference—the surface-air temperature anomaly—is somewhat too small. Hence, one can argue that the regional-temperature anomaly in figure 1 may be conservatively constructed, because differencing the decadal fields in this fashion would suppress the temperature change by not accounting for urban heat-island effects. These considerations suggest that more ingenious processing of existing records in the quest for climate analogs is warranted. The recent instrumental record, however, will not yield analogs associated with increased average surface-air temperature that are as high as those currently estimated, 1.5° to 4.5° C for 2 × CO_2 obtained from the GCMs (NRC 1983). Simulation modeling appears to be the only pathway to more reliable regional estimates of future climate states corresponding to a CO_2 doubling or its equivalent if the several other greenhouse gases are included.

Climate Scenario for 2 × CO_2 Period (2030–2070)
Increase in the annual average surface temperature = 2° to 4° C (line B1, table 1). The several premier GCM simulation models widely accepted in the U.S., but still in need of several specific modeling improvements, project that the average NH temperature could be 2° to 6° C warmer when at climatic equilibrium and when 2 × CO_2 atmospheric concentrations are achieved and hypothetically held constant for another 50 to 150 years. The type of "computer climate construct" has its conceptual limitations in that the real global system may well be behaving in a transient, nonequilibrium manner (MacCracken and Luther 1985). The variation in simulation results (noted above) would suggest a factor-of-two uncertainty in the magnitude of the projected temperature change. The uncertainty on the timing of the effect is reflected in the bounds of the window, 2030–2070; if, however, the several other greenhouse gases are included in the timing considerations, then such warming effects could occur as early as 2030. The estimate of 2030 as the earliest arrival of a 2 × CO_2 world does not take into account the prospects for any remedial or preventive actions by the world's nations, or the recently published information regarding the temperature-dependency of natural methane emissions (Wuebbles et al. 1988). The conservative range of 2° to 4° C reflects some judgmental correction for the transient climate response to lag the equilibrium-state projection.

Change in precipitation: + 10 percent globally, and ± 20 percent in California (line B2, table 1). The GCM models indicate that the hydrological cycle on the global scale could be enhanced in the 2 × CO_2 world by about 10 ± 5 percent. On the regional scale (i.e., making a grid point-by-grid point comparison), the GCMs show marked disagreement in their predictions for changes in precipitation, with disagreement concerning summertime precipitation being the greatest. For California, GCM projections of changes in precipitation should be considered highly questionable until more realistic topography and oceanic couplings are included in the simulations. Hence, we recommend that impact studies on watersheds consider the range of projections for annual precipitation spanning the range -20 to 20 percent, to facilitate examination of infrastructure sensitivities to the climate scenario.

Rise in mean sea level of 0.2 to 1.0 meters (line B3, table 1). Global sea level has risen perhaps $0.05-0.2$ m over the past century; in California (and elsewhere), rates are also affected by the upward and downward movement of the Earth's crust due to tectonic movements. Our estimate of the rise in mean sea level for the mid-twenty-first century is drawn from NRC (1989) and represents the range between their low estimate (.2m) and the high estimate (1.0m). Since 1987, MacCracken (1989, in preparation) has reexamined this issue using different, but very simple, ocean models of diffusion, upwelling, and bottom-water generation. The inclusion of processes that can explicitly represent rate of southward advance of tracers in the newly formed bottom water in the North Atlantic suggests slower rates of sea-level rise. Hence, the most likely increase in mean sea level for the mid-twenty-first century is probably toward the lower bound of 0.2 meters.

Rise in snow level of 100 meters for each 1° C increase in temperature (line B4, table 1). The above estimate of rise in snow level of 100 meters for each 1° C temperature rise is a reasonable assumption and is that used in the national assessment of effects of global warming on Australia (Pearman 1988). Since, during California spring melting periods, most of the melt occurs during warm afternoons on the heated slopes, the ambient-air lapse rate would be about 1° C per 100 meters. Exceptions could operate to make this estimated rise in snow level too small; one exception involves the possible increased frequency of warm winter storm systems from the southwest, in which case warm rain could erode the previously deposited snow, and, second, if the snow line rises to be within the storm system during deposition, then the snow line could rise about 150 meters for each 1° C rise in temperature. Both of these latter effects would operate during the winter season to enhance winter runoff.

Air pollution: Increase in the peak surface-air concentrations of ozone downwind of urban areas (line B5, table 1). Penner et al. (1988) have investigated the connections between global climate change and regional air pollution; impacts include the feedback of globally altered conditions (e.g., global background of ozone and enhanced UV_b from global ozone depletion) on the regional processes of temperature-dependent emissions from soil and vegetation and temperature-dependent reactions rates. Figure 3 presents the schematic of these connections, which lead to the potential for enhanced surface-air peak ozone concentrations in the warmer world of $2 \times CO_2$. The estimates of enhanced peaks (e.g.,

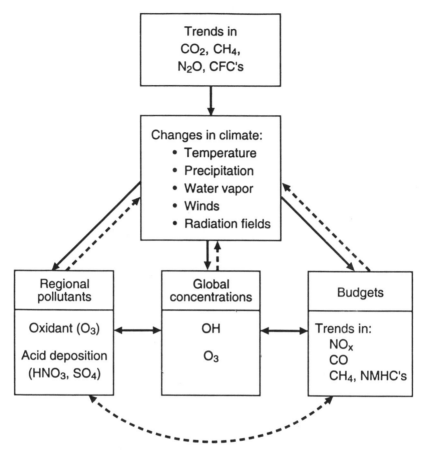

Figure 3. Processes interconnecting climate and global and regional chemistry.

+10 to 20 percent) do not take into account increasing population, altered traffic patterns, changes in abatement procedures, and so on. Their efforts to date have been in the spirit of a sensitivity study of ozone air-pollution potential in a warmer world, to identify air pollution and changes in acid-rain formation processes in the projected climate setting of the twenty-first century. Figure 4 from Penner et al. (1988) illustrates the nature of this sensitivity. Uncertainties in the analysis are probably comparable to the inherent uncertainty of a factor of two in the global surface-air temperature projected for the mid-twenty-first century.

The UV$_b$ to the Earth's surface: Increased by 50 percent (line B6, table 1). Wuebbles (1989, private communication) has estimated that the CFC emissions to date will result in a 25 percent depletion of stratospheric ozone at mid-latitudes on the global scale in the mid-twenty-first century.

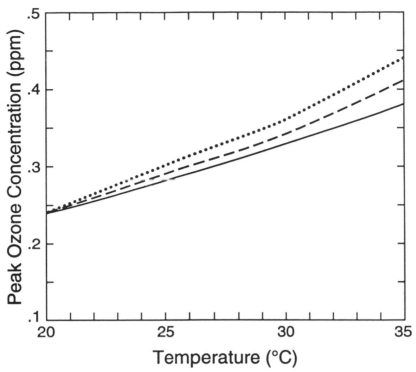

Figure 4. The effect of changes in temperature and biogenic hydrocarbon emission on the peak O_3 concentrations predicted in a box model calculation of urban O_3 formation. The solid line shows the effect of temperature alone. The short-dashed curve shows the increase expected when the flux of biogenic hydrocarbons is increased along with the temperature. The long-dashed curve shows the change expected if the initial concentration of biogenic hydrocarbons is increased in proportion to the expected increase in the flux of biogenic emissions.

The projection of an increased UV_b flux of 50 percent is consistent with this value of estimated ozone reduction; this projection was incorporated into the previously cited study by Penner et al. (1988). The projection of health effects on humans within urban areas is complicated by the fact that the urban tropospheric ozone increase will moderate the increased UV_b flux from above. The net increase in UV_b from these compensating effects would need to be estimated. In rural areas not affected by urban plumes, the potential increase in frequency of skin cancer from the enhanced UV_b flux can be estimated as about twice the percentage change in UV_b flux, assuming other life-style factors remain constant (NRC 1984). Agricultural crops and natural ecosystems would also be subjected to increased water stress, CO_2 enrichment processes, enhanced ozone peaks, and the higher UV_b fluxes. The effects of this combination of stresses remain to be explored for primary crop types through experimental studies in chambers and by other means.

Storm tracks (line B8, table 1). In the Australian national assessment of greenhouse warming (Pearman 1988), the climate scenario includes a poleward shift of climate zones and a consistent poleward movement of tropical storm tracks. In our scenario for the mid-twenty-first century, it is entirely reasonable to project a 100- to perhaps 200-mile poleward shift of major climate zones. Such shifts are usually estimated on the basis of the latitudinal shift required to match elevated mean seasonal temperatures consistent with those projected on a global scale for a few degrees of warming. In mountainous areas, such extrapolations are misleading, in that climate zones may in fact stay relatively fixed and migrate mainly upslope. California, hence, seems to be a mixture whose precise mosaic or pattern of change is not easily identified by simple rules. By mapping the seasonal mean temperature patterns existing now, and moving boundaries along temperature gradients appropriately, one may be able to provide maps of new climate zones consistent with our climate scenario and the complexities of local topographic relief.

The potential shifts of storm tracks, both tropical and those in the north Pacific, will have significance for the state's activities. The drier autumns projected for the period 2000–2010 suggest a northward shift of extratropical storm systems, accompanied by perhaps a similar shift in the eastern Pacific hurricane paths in western Mexico. Such subtle changes in storm tracks are (author's opinion) impossible to confirm or validate from the results of the present coarsely zoned climate simulation models. The potential shifts that can be imagined as plausible in the mid-twenty-first century (e.g., say two degrees of latitude) cannot be resolved with any accuracy with the five-degree grid-point resolution now available. Such concepts remain a subject for future climate research.

In 1988 the California state legislature mandated a study of global warming, encompassing a statement of the problem, the spectrum of views about the greenhouse effect, and assessments of impacts for the state of California. The California Energy Commission was named to coordinate the effort with major inputs from the various state agencies and departments. A preliminary study draft was released in June 1989, and the University of California workshop organizers made copies of this study available to workshop participants. The climate scenario (table 2)

TABLE 2. State Study Scenario for Global Warming

1. Effective CO_2 doubling by the year 2030.
2. Temperature increase of 3.0 degrees centigrade.
3. Rainfall: 0 percent change in annual precipitation.
4. Ocean level rise of 3–3½ feet (one meter).
5. No change in current economic and social trends.

on which the state study was based was provided by Mr. Allen Edwards of the CEC; it is presented for comparison with our two workshop scenarios.

REFLECTIONS ON THE WORKSHOP PROCESS

During the several months of workshop organization and preparation it became abundantly clear that our UC community of researchers held quite divergent views regarding the evidence of global warming, and some tension emerged between those of differing opinions. In the spirit of inclusion and for openness of dialogue, the following statement of the workshop chair served to place our process in perspective.

As we begin our UC Workshop process, we wish to be forthright in informing all of the participants that at the moment there is a great diversity of views about the greenhouse effect, its evidence as reflected in the observational record, and the seriousness with which society should consider the greenhouse issue. A vigorous pro-greenhouse view (Kellogg 1988) holds that "a strong case can be made for the assertion that we are in the grips of an irreversible climate change." One alternative view holds that the land-based surface-air temperature records, frequently used to document global warming, are flawed by man's activities and the process of urbanization; further, that rural stations in the U.S. do not provide evidence of global warming—in fact, in some analyses the analyst claims cooling rather than evidence of warming. Another view holds that cloud processes and their effects on the radiation balance of the atmosphere are very poorly represented, if not incompletely represented, in the climate models; one possibility, for example, is that clouds in the "warmer world" will contain more liquid water than now, hence, these clouds would reflect more incoming solar radiation back to space, and thus significantly moderate the warming of the underlying atmosphere. Still another view holds that ice ages have occurred about every ten thousand years or so on the planet, and that the so-called greenhouse warming, if real, could turn out to be a blessing in helping to offset the potential cooling of such a future ice age. These are only a few of the current viewpoints about the greenhouse issue. Since the very warm summer of 1988, when political attention became quickly and intensely focused on the issue, both the pro-greenhouse view and the alternative views have gained more visibility in the media. It is clear that scientific differences exist, that views widely differ, and that new and improved simulations of future climate states corresponding to a doubling of the CO_2 concentration are at least several years away.

In a sense those who hold divergent views about the greenhouse issue are asking the climate modelers, and others, very difficult questions. If they were easy questions to answer, the subissues might be readily resolved. By the asking of these "hard questions" and identifying analysis weaknesses, the divergent view holders become the teachers of those in the consensus. By working to resolve these important subissues, we will together arrive at an improved understanding of the greenhouse issue and its effects on society in the twenty-first century. The organizers of the UC workshops believe that our workshop process can and should contribute to this most worthy goal.

We also wish to restress the uncertainty of the assessment results and findings, which depend in no small part on the proposed climate scenarios constructed, and the infrastructure assumptions. Our proposed climate scenarios for 2000 to 2010 and 2030 to 2070 are not presented as precise predictions of future climatic conditions for California, but rather as plausible estimates of potential climate change which have been made as consistent as possible with the current consensus understanding of the greenhouse effect. The climate scenarios are but tools to assist in the examination of the sensitivities and vulnerabilities of our infrastructure to climate change. Having learned of and become aware of these sensitivities and vulnerabilities, it is our hope that society will gain awareness, understanding, and knowledge so that planning for future energy supplies, water resources, and the agricultural/industrial sectors could profit from becoming more robust in the face of natural climate variability and potential climate changes in the future.

We also note that there exists in some quarters the feeling that the evidence of the greenhouse warming is so uncertain and the timing of the effect so unknown that it is premature to take action. Further, some hold that perhaps it is prudent to wait until the evidence and the projections are improved or until new observations providing more conclusive evidence become available. In 1983 this cautious approach to immediate policy actions prevailed in the U.S., and the years of 1980–1988 were marked by recognition of the uncertainties and the attitude of waiting for improved computer simulations and understanding, or more definitive observational evidence. However, during the 1980s there was a cascade of environmental events that led to a rapid growth of awareness of the greenhouse issue. This cascade of events is well known to all of us: the discovery of the ozone hole over Antarctica, the reactor accident in Chernobyl (whose airborne unique radioisotopic signature was transmitted around the world), the weather extremes of the summer of 1988, and the chemical priming of the arctic polar vortex as another ozone hole. These events have dramatized the fact that global environmental changes can happen quickly, involve all nations, and that there are surely surprises hidden in the midst of growing understanding of the potential for global change. In 1988 and now again in 1989, legislation is being proposed in Congress pertaining to the development of a national energy policy and to the protection of our global climate resource. In a sense we are committed to a quest to improve our understanding of global change, to enhance our simulation models so that the future responses of the coupled atmospheric-ocean system to various greenhouse gas emission profiles can be more faithfully simulated, and to identify those plausible policy actions that can be justified for reasons in addition to their "greenhouse" effect. Examples of such policy actions need to be identified, actions which would be reasonable to implement in any case for reasons of conservation and efficiency, that would significantly help to delay or blunt the potential greenhouse impact of the next century. Many feel that our country may well become party to an international convention to reduce greenhouse gas emissions by 1995 or even perhaps before. In a very real sense we have no other choice than to move forward, despite the existing controversy, toward the needed improvements in understanding and awareness, enhanced assessment methods, and the identification of prudent potential policy actions.

MAJOR PANEL FINDINGS

Water

In the event of projected global warming, the entire range of water problems that California faces today will be intensified, even if precipitation levels remain unchanged. In the absence of an effective response, major disruptions could occur in California's surface-water supply systems. Since there is much uncertainty both about the dimensions of global warming and the impacts of such warming on California watersheds, an immediate and prudent response would place high priority on solving today's water problems. Such a response would include:

1) building more flexibility into the existing water allocation and delivery system,
2) developing incentives that will promote economically efficient levels of water use, and
3) developing institutions to manage groundwater quantities and qualities effectively.

Simultaneously, it will be important to monitor for signs of climate change. This should involve a continuation of efforts to develop baseline information on climatic, hydrologic, and ecological trends. Additionally, efforts should be made to improve facilities for maintaining, analyzing, and distributing these data.

Agriculture

The major goal of the agricultural sector in response to the challenge of global warming should be to maintain and enhance, if possible, the resilience of agriculture systems to climate variability and change. Attention should be given to the control of possible pest population explosions under altered climatic conditions—for example, in the cotton crops of the southern San Joaquin Valley. Flexibility, based on genetic modification of crops, should be developed to aid in the adaptation of crops to different climatic or growing conditions. The projected warmer climate with a doubling of CO_2 projected for the twenty-first century means that agricultural biomass may increase by as much as 50 percent (assuming that water for increased production is available). Hence, the possible use of biomass residuals as fuel for energy generation, displacing the use of fossil fuels, should be seriously examined.

In addition, agroforestry should be pursued for its potential for sequestering carbon. Since such agroforests might displace agriculture and compete with needs for cropland, such measures may provide only a means of temporarily delaying greenhouse effects. Afforestation, in general, requires research to determine which tree species do not add to the present air pollution problems that are expected to be worsened in a warmer world, especially with the presence of increased population.

Natural Ecosystems

Very little knowledge now exists on how rapidly tree and plant species could migrate under changing climatic conditions. The panel estimates that 20 to 50 percent of the area occupied by natural ecosystems will no longer be suitable for the communities that exist there now. Diebacks, inadequate ground cover, enhanced erosion, and loss of species could well prevail and contribute to the worsening of adverse effects in other sectors such as water resources.

The possibility of a strengthened California ocean current developing in response to global-scale warming emerged as a most significant finding in our workshop process. The potential implications of a strengthened California Current are considerable: the California coastline could have cooler, foggier summers; the onshore sea breeze phenomena could be strengthened, increasing the wind energy potential on coastal ridges and confining valleys; and conceivably, some fish species might find these cooler summer waters a haven in comparison with previous projections. This finding emphasizes the strong need for more research on mesoscale response of the coupled ocean atmospheric system to large-scale, greenhouse-forced warming.

Human Dimensions

The prediction of future climate states of the atmosphere has historically been an uncertain business. Despite noble efforts, incomplete climate simulation models and inadequate databases will persist beyond the year 2000. Although uncertainty seems destined to persist, some key uncertainties should be reduced by improved understanding. How individuals and institutions behave and make strategic decisions in the face of uncertainty remains a key issue. The public is believed to be more sensitive to sequences of climatic extremes than to gradual changes in the mean state of the atmosphere on regional scales. Climate modelers and social scientists need to take this characteristic much more into account in their cross-discipline studies of human perception of global change. Research is needed into the question of public response and behavior of the consumer when presented with choices and multiple policy opinions.

The panel recommends the development of a socioeconomic model for the state of California in order to explore the impact of various policies or sets of policies aimed at delaying, mitigating, or preventing greenhouse effects. The potential for there to be "greenhouse winners" and "greenhouse losers" within the state of California could well exist. Political competition between such groups could make the adoption and implementation of policy more difficult than it would be otherwise.

Climate Caucus

The Climate Caucus participation in our workshop process served several important functions. The first was to serve as a resource to help the separate panels interpret the projected climate scenarios for use in the impact studies and their discussions. The second was for the caucus to become better informed about the interactions and the informational needs of the impact assessors dealing with natural ecosystems, water resources, agriculture, and human systems; this is a necessary step toward providing more complete climate scenarios in the future. The third purpose was to develop plans and recommendations for extending collaborative efforts between researchers and those studying impacts of climate change on California. The most salient recommendations and research areas are highlighted below.

1) In the near term, existing climate data for our current control climate and the $2 \times CO_2$ simulations could and should be made available to impact modelers with much higher temporal resolution, even though it remains on a coarse spatial grid for the present. Specific questions, such as storm speed, frequency changes, and intensity changes, could be addressed with this improved information.

2) Given the spatial and climatic complexity of California, it is essential to develop the means for providing high-resolution spatial information on climatic variables needed for use in the evaluation of potential impacts on water, agriculture, and natural ecosystems. Within the UC system, such research is being initiated and needs to be continued with accelerated support.

3) During the workshop a number of mesoscale responses to global climate change, or greenhouse forcing, were discussed; most notably, the possibility of interactions between the much warmer land and the California ocean current was discussed. The projection was made that in the $2 \times CO_2$ world the California Current would be enhanced, cooler inshore seawater temperatures would arise through stronger upwelling, and the resulting enhanced land-sea thermal contrast could well result in cooler summer climates along the coast and stronger wind energy

resources during this season of the year. This example illustrates the possible complexity of the climate response of a coupled atmosphere-ocean system to a larger-scale forcing. The investigation and development of nested mesoscale simulations of coupled systems is an essential research requirement. The resources of the UC system can and should be focused on this need and opportunity.

4) It is clear that, even if only a few of the research needs identified by the Climate Caucus were pursued, additional computer resources would need to be provided to facilitate the use of high-resolution temperate-climate data and the use of high-resolution climate scenarios (when they become available) and to pursue the simulation of the mesoscale climatic response of a coupled atmosphere and ocean to large-scale greenhouse forcing. The existing research centers, as well as new MRUs (multi-campus research units) that may appear, will need to be linked.

Energy Caucus

Members of the Energy Caucus were dispersed into the four primary panels of the workshop for the purpose of identifying and summarizing the most significant impacts of global climate change in California on energy supply and demands. In the course of this process, the Energy Caucus developed recommendation or action items reinforcing the findings of the panels on agriculture, water resources, natural ecosystems, and human dimensions. The Energy Caucus findings are summarized below:

1) Increase water conservation efforts. Do more with less, that is, use available water more effectively in all infrastructure components—urban, agricultural, and industrial.
2) Construct more water storage facilities, some of which could also serve as impoundments for hydroelectric facilities and enhance carriage water availability.
3) Pursue research to evaluate opportunities for renewable biofuels to replace, to whatever extent possible, the nonrenewable fossil fuels.
4) Gather information to evaluate future energy costs of tillage, transport, water pumping, and the processing/preservation of perishable foods.
5) Pursue research about species of trees most appropriate for the sequestering of carbon which do not adversely increase natural or biogenic emissions to the atmosphere.
6) Conduct comprehensive studies concerning human response to multiple policy measures designed to reduce overall use of fossil fuels and promote energy conservation.

CONCLUSIONS

Beyond the summary findings and recommendations just reported, many of the workshop participants returned to their research and teaching with a strong urge to complete the quest for a rational, scientifically sound response to the challenge of global change. The chairmen of both the climate and energy caucuses have articulated within this volume their distilled insights. It is abundantly clear that a reasonable course of action must be grounded on the best currently available science, and that a strategic research agenda must improve and enhance our understanding of global climate change and provide more reliable and detailed regional assessments to guide decision makers. It is also true that the uncertainties in the projected effects are uncomfortably large and in need of reduction through focused research and development of advanced climate simulation models.

The scientific understanding of planetary warming and improved assessments of climate change on the regional scale (e.g., regions like California) are very high priorities for federally sponsored research. However, as uncertain as these projections may be, nature has sent and is in a sense sending us a warning signal that human activity and technology are modifying our planet, and the potential for dramatic alteration of our climate system is present.

How do we respond to the greenhouse challenge in view of the apparent uncertainties? The Edison Electric Institute has advised the electric utility industry to begin to factor global climate change into its planning process, particularly in service regions where there is a large air-conditioning load or dependence on hydroelectric generation. This recommendation is consistent with our UC/DOE workshop findings that our infrastructure sensitivities and vulnerabilities to global change should be identified and that infrastructure resilience to global change should be promoted in reasonable ways. We must not be paralyzed by scientific uncertainty from pursuing reasonable actions in management of energy, water resources, agriculture, urban environments, and the resource base represented by our unmanaged ecosystems.

Admiral Watkins, the current secretary of energy, has stated that our nation's use of our cheapest source of energy is being threatened by global and environmental concerns; he has called for a new national energy strategy to be developed; studies have been mandated and are under way to ascertain to what extent energy efficiency and conservation can be utilized and promoted to achieve a reduction in U.S. emissions of CO_2. The new national energy strategy is expected to be made public in 1990 for comment. It is appropriate here to offer some suggestions specific to California on a reasonable policy approach to the greenhouse challenge in view of our considerable gaps in knowledge.

Let us define a reasonable policy approach to the "greenhouse issue" as the pursuit of actions that reduce the greenhouse threat but are justified on other grounds as well, such as national energy security, the improvement of local environmental conditions, or U.S. competitiveness—for example, increased GNP per unit of energy used. In a sense, decision makers should momentarily forget about greenhouse impacts and address those policy actions that serve general national goals. Policies that serve several purposes should be favored. Most such proposed actions would find support from both pro-greenhouse and anti-greenhouse factions. Simply stated, it makes good sense both to generate energy more efficiently on the supply side and to use conservation to reduce the demand side. Meanwhile, such actions can serve as a low-cost or no-cost insurance policy against projected future global warming. An illustrative list of reasonable policy actions is given below for analysis by appropriate studies of applicability and cost in California. The list is no doubt incomplete, but it is illustrative; it contains suggestions that go beyond the electric utility industry because important trade-offs may well exist between different economic sectors.

- Improve power plant efficiency, particularly in new facilities;
- Improve energy efficiency standards for new major buildings;
- Set conservation standards for new and old residential buildings (e.g., the newly initiated program of New England Electric for conservation programs, utility-paid, in a demonstration town of 60,000);
- Set more rigorous appliance standards;
- Promote higher mpg automobile options;
- Support promising alternative energy options (e.g., photovoltaics, modern gasified bombast plants, and improved gas turbines, following the findings of the second UC/DOE workshop on global warming, September 1989);
- Initiate mandatory water conservation programs;
- Require drought-resistant landscaping for new residences;
- Require clean fuels for the transportation sector.

Clearly the items on this list are not new. What may be new is the concept of developing a low-cost insurance policy against potential disruptive changes in the climate system projected for the future. From this illustrative list and its expansion to an even larger list, one could very well design a reasonable set of policy options, justified in their own right, but indeed serving as insurance against the emerging threat of climate change—whether induced by the vagaries of nature or by human activities and population pressures.

Our global system is an invaluable and irreplaceable resource. California has an opportunity to lead the way toward responsible stewardship of our global environmental resources. Let us not lose that opportunity in the paralysis of debate.

REFERENCES

Edmonds, J. A., et al. 1986. *Future atmospheric carbon dioxide scenarios and limitation strategies*. Park Ridge, N.J.: Noyes Publications.

Engelbert, E. A., and A. F. Scheuring, eds. 1982. *Competition for California water: Alternative resolutions*. Berkeley, Los Angeles, London: University of California Press.

Karl, T., and P. D. Jones. 1989. Urban bias in area-averaged surface air temperature trends. *Bulletin Amer. Meteor. Soc.* 70, no. 3.

Kellogg, W. W. 1988. Climate change is already here to stay. *EOS*, vol. 69, no. 44.

MacCracken, M. C. 1989. Private communication/publication in process.

MacCracken, M. C., and F. M. Luther. 1985. *The potential climatic effects of increasing carbon dioxide*. UC-DOE Report ER-0237, December. Livermore, Calif.

National Research Council (NRC). 1983. *Changing climate*. Carbon Dioxide Assessment Committee. Washington, D.C.: National Academy Press.

―――. 1984. *Causes and effects of changes in stratospheric ozone: Update 1983*. Washington, D.C.: National Academy Press.

―――. 1987. *Responding to changes in sea level: Engineering implications*. Washington, D.C.: National Academy Press.

―――. 1989. *Global change and our common future: Papers from a forum*. Washington, D.C.: National Academy Press.

Pearman, G. I., ed. 1988. *Greenhouse: Planning for climate change*. Melbourne: CSIRO Publications.

Penner, J. E., et al. 1988. *Climate change and its interactions with air chemistry: Perspectives and research needs*. UCRL-2111, June. Livermore, Calif.

Smith, J. B., and D. A. Tirpak, eds. 1989. California. In *The potential effects of global climate change on the United States*, chap. 14 (251–285). EPA-230-05-89-'050. Washington, D.C.: U.S. Environmental Protection Agency.

Wigley, T. L. M. 1986. In *Future atmospheric carbon dioxide scenarios and limitation strategies*, by J. A. Edmonds et al. Park Ridge, N.J.: Noyes Publications.

Wuebbles, D. J. 1989. Private communication.

Wuebbles, D. J., et al. 1988. *The role of atmospheric chemistry in climate change*. UCRL-97811 Rev. 1, November. Livermore, Calif.

TWO

Greenhouse Gases:
Changing the Global Climate

Michael C. MacCracken

Observations of the various temperatures of nearby planets and geologic reconstructions of climates tens to hundreds of millions of years ago, in association with calculations of the visible and infrared radiative fluxes, clearly demonstrate the potential for significant climatic change as emissions from societal activities alter the composition of Earth's atmosphere. In seeking to estimate the extent of future change, we can gain general guidance from analogs drawn from study of past climates, from analytic and laboratory studies, and from the trends beginning to emerge from the recent record, but none of these approaches can provide reliable, highly resolved estimates of the complex and unprecedented changes that society has initiated. There is also no definitive and convincing measurement—except for awaiting the outcome of our great "geophysical experiment"—that can alone serve as the basis for predicting the future climate.

In the absence of such traditional approaches to addressing the coupled physics and chemistry questions posed by the complex atmosphere-ocean-land-biosphere system, we are forced to rely on development of numerical models that seek to emulate all of the important and interacting processes.[1] The most comprehensive of these models are known as general circulation models (GCMs), which attempt to represent the three-dimensional, time-dependent character of the atmosphere and/or oceans. Modeling of the global climate is a particularly difficult challenge because the time scales of interest vary from hours to centuries, and spatial scales from kilometers to global. Even these broad scales, however, do not encompass the complete range of scales of atmospheric and oceanic motions—so that even if we can model the scales of interest, we must parameterize the effects of the smaller and shorter scales. Thus, we are forced into constructing models that not only incorporate the limitations in our understanding of large-scale processes but also approximate the effects of smaller-scale processes known to be important in a deterministic, if perhaps not in a statistical, sense.

It is an open question whether such theoretical constructs can provide a sufficiently convincing basis for implementing important policy decisions concerning fundamental aspects of societal development and living styles. Although it is not possible to establish that the models we construct are correct—until after the fact—it is essential, as a minimum, that the ability of the models to represent past and present climates be clearly demonstrated against a range of past and present climatic changes if the models are to be used as a basis for policy formulation.

PREDICTING GLOBAL-SCALE WARMING

The potential global average warming expected to result from an instantaneous and perpetual doubling of the atmospheric carbon dioxide concentration (or its radiative equivalent from increases in the concentrations of carbon dioxide, methane, nitrous oxide, chlorofluorocarbons, and other greenhouse gases) is a convenient measure of the climate's sensitivity to radiative perturbations. In such calculations, it is assumed that oceans need to be represented only in terms of the heat capacity of their upper (50 to 100 m), mixed layer, which mainly governs the seasonal cycle thermal inertia, rather than in their full complexity. This simplification dramatically reduces the computer time needed to achieve a new statistical-equilibrium climate. Although these equilibrium calculations are not expected to simulate realistically the climatic response to the steadily changing composition of the atmosphere, especially because they do not account for potential changes in ocean circulation, they do indicate the commitment that society is creating to future climatic change. In addition, projections of greenhouse gas emissions suggest that we may become committed to these climatic changes by some time around the middle of the next century (and are already committed to about half of the projected changes even if we could somehow completely stabilize the atmospheric composition at present concentrations).

Results from models developed somewhat independently at five leading climate-modeling centers (i.e., National Center for Atmospheric Research [NCAR], NOAA Geophysical Fluid Dynamics Laboratory [GFDL], NASA Goddard Institute for Space Studies [GISS], Oregon State University, and the United Kingdom Meteorological Office),[2] when used to make roughly comparable simulations, estimate that global average surface-air temperature will increase by about 3° to 5° C for a doubling of the CO_2 concentration.[3] These values tend to be in the upper half of the estimated sensitivity range of 1.5° to 4.5° C adopted by National Research Council panels starting about ten years ago.[4] Model intercomparison studies suggest that the treatment of clouds is the major cause of differences among the models.[5] Variations in cloud properties, which are not now well treated, could also be important factors in altering the sensitivity estimates.[6]

There are several reasons to suggest that this order of magnitude is roughly correct. A doubling of the CO_2 concentration results in an increase in the net tropospheric-surface trapping of infrared radiation by about 4 to 5 W/m^2. A number of tests of the models have been carried out that examine model responses to a wide array of other perturbations, which help provide some assurance that models are behaving realistically. The models represent the seasonal variations in mid-latitude temperature, which are driven by solar radiation variations of ± 100 W/m^2 about the mean, to within about 10 percent.[7] Several models have successfully simulated the evolution of climatic conditions since the last glacial maximum, which involved seasonal changes in mid-latitude solar insolation of up to ± 40 W/m^2 as a result of changes in Earth's orbit.[8] The models also do not exhibit an excessive sensitivity to the transitory perturbations of a few W/m^2 caused by major volcanic aerosol injections, in agreement with the marginally detectable observed changes. Most recently, model results seem to be representing characteristics of the low-frequency climatic oscillations evidenced by the Southern Oscillation, the Quasi-biennial Oscillation, and other natural features.[9]

Thus, although various processes in the models may not yet be fully and adequately represented, the model results provide strong evidence that the climatic sensitivity to a CO_2 doubling will be a few degrees, not tenths of a degree or ten degrees.[10] Soviet reconstructions of climatic conditions tens of millions of years ago, when the natural CO_2 concentration is believed to have been two to five times current levels, also suggest such a sensitivity.[11]

PREDICTING REGIONAL-SCALE CHANGES

Given the ranges of local temperature variations, a global-scale warming of a few degrees may seem a small price to pay for the benefits of energy, agriculture, refrigeration, and transportation. There is thus interest in determining the prospective changes with more spatial and temporal detail in order to be better able to evaluate potential impacts.

All model results indicate that the warming will be somewhat greater in high latitudes as the freeze season shortens and the warm season lengthens, the warming being amplified as the insulating effect of sea ice is reduced.[12] Paleoclimatic studies also indicate greater temperature changes in high latitudes than in low latitudes.

On finer scales, the limitations of present models make estimation of regional scale climatic changes problematic. The horizontal resolution of currently available general circulation models is typically five degrees of latitude and longitude (roughly 550 kilometers at the equator). In mid-latitudes, this resolution provides one grid point (with one value of temperature, wind speed, precipitation, etc.) for an area roughly the size of Colorado; in northern California, one grid point represents the diverse region extending from the Pacific Ocean west of San Francisco across the San Joaquin Valley and Sierra Nevada mountains into the deserts of Nevada.

Given such coarse resolution, developing an observation set against which to compare is not straightforward. Despite resulting limitations, when point-by-point comparisons are attempted, present models appear able to represent wintertime temperatures better than summertime, and larger scales better than smaller scales.[13] Over areas the size of the United States, differences between model results and observations of seasonal average temperatures are typically several degrees in summer, with deviations reaching up to 10° C in areas where the timing and extent of summertime drying are significant factors in determining surface temperature (e.g., the Midwest). Not only are treatments of hydrology in the models highly simplified (e.g., use of a fifteen-centimeter-deep "bucket" to represent soil moisture), but other assumptions in some models also make prediction of summer temperatures very difficult (e.g., no treatment of the diurnal cycle, poor treatment of nonraining clouds).

Intermodel comparison of the estimates of the sensitivity of regional climatic conditions to a doubled CO_2 concentration show very little spatial coherence, especially in summer and over land areas (see fig. 1). Although agreement among models does not assure that their results are correct, the low correlation coefficients among results from different models indicate that there is as yet no skill in predicting whether changes will be larger, for example, in the Southeast or in the Southwest. About all that model results indicate on a regional scale is that the local variations about the global average temperature changes will be strongly influenced by the balance between changes in precipitation (which will increase in many areas) and evaporation (which will almost certainly increase everywhere). Because models do not yet include comprehensive representations of the hydrologic cycle, however, the more or less random variation of results about the global average provides little useful indication of how precipitation may change. Improving this situation is a critical research objective.

Refining the horizontal resolution of the climate models is also expected to be necessary to improve simulation of regional climatic conditions significantly. There is a practical difficulty in accomplishing this, however; each halving of the grid size requires about an order of magnitude increase in computer time (four times as many horizontal grid

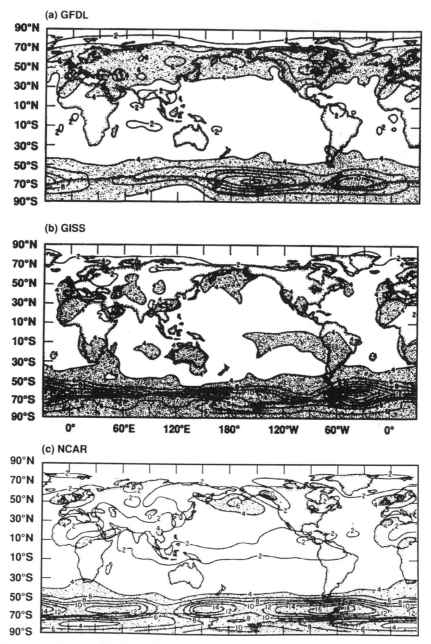

Figure 1. Geographical distribution of the surface-air temperature change (°C), $2 \times CO_2 - 1 \times CO_2$, for June-July-August simulated with: (top) the GFDL GCM by Manabe and Wetherald (1987); (middle) the GISS GCM model by Hansen et al. (1984); (bottom) the NCAR GCM by Washington and Meehl (1984). Stipple indicates temperature increases larger than 4° C.

points, twice the number of vertical grid points, and twice the number of time steps owing to the Courant instability condition, less not having to calculate some processes twice as often). Thus, fifty-year model simulations that now require a few hundred hours of supercomputer time would require a few thousand hours—making difficult significant testing of model parameterizations. A variety of other approaches to improving model resolution also merit consideration, including: refinement of the grid only in critical areas; driving a finer-grid mesoscale model with boundary conditions derived from a coarser-grid global model; using empirically derived relationships to go from large- to small-scale conditions (as is done in weather forecasting); or, perhaps, using more efficient computational techniques.

In addition to the computer demands required for greater resolution, a number of other factors also pose demands for increased computer time. Adding interactive chemistry to treat adequately the many interactions posed by the increases in chemically active trace gases could increase the number of prognostic equations from five to a few dozen.[14] Interactively coupling atmosphere and ocean models, especially ocean models that have the fine resolution (e.g., 0.25° or finer) needed to resolve the important eddy motions and that represent the long time constants of the deep ocean, can increase computational requirements many times. Longer simulations are needed in order to treat cases with slowly increasing greenhouse gas concentrations. More accurately representing complex processes, such as hydrology, convection, and the land biosphere, will further increase computer demands.

Another troubling feature of models concerns their ability to represent the full spectrum of variability seen in observations. Until recently it has appeared that model behavior is relatively stable, changing only slowly in response to external forcings, especially when the models do not fully couple the ocean to the atmosphere. There are, however, indications that past climatic conditions have changed relatively rapidly. For example, about a 0.3° C Northern Hemisphere warming occurred several times in the five years around 1920, and relatively sharp changes have occurred several times in the eighteen thousand years since the last glacial maximum. In the last several years, a few model calculations have exhibited either relatively rapid (i.e., decadal-scale) fluctuations, or even multiple equilibria.[15] Inquiries into the potential for such surprise climatic shifts as well as into changes in the frequencies of short-term extreme events deserve greater attention.

PROJECTING TIME-DEPENDENT CLIMATE CHANGE

Actual calculation of the rate at which climate should be changing requires a comprehensive atmosphere-ocean model that includes consideration of the climatic effects of volcanic eruptions, solar variations, and the increasing concentrations of greenhouse gases since the preindustrial period and then into the future. There are not yet adequate models nor sufficient information fully to drive such models, although initial attempts are being made. For example, a calculation done at GISS using a simplified representation of the deep ocean, starting in the 1960s, shows a gradual warming that leads to global average temperatures exceeding maximum temperatures of the last interglacial period within the next few decades, depending on scenario assumptions about future changes in emissions.[16]

In lieu of complete calculations, an interpolation technique has been used to look at the consistency of model estimates of climate sensitivity and recent climatic change. Assuming the climate sensitivity to a doubling of the CO_2 concentration (or equivalent through a radiative contribution by the several greenhouse gases) is actually a few degrees, then we would expect to observe a response as a consequence of the 25 percent increase in CO_2, the doubling of the CH_4 concentration, and the increases in concentration of other trace gases above preindustrial levels. Dickinson and Cicerone (1986) estimate that the flux change from these combined changes is about 2.2 W/m^2, just half the 4.4 W/m^2 they estimate would result from a doubling of the CO_2 concentration. For a climate sensitivity range of 1.5° to 4.5° C for a CO_2 doubling, this converts to a commitment to a temperature increase, *at equilibrium*, of about 0.75° to 2.25° C, based on the present atmospheric composition. There is currently considerable disagreement about the lag behind equilibrium caused by the time it takes to warm the oceans, with estimates from simple ocean models of the response time of the oceans ranging from a few decades to more than a century.[17] Accounting for this lag effect, the warming over the past 150 years would be expected to be perhaps 0.4° to 1.5° C.

Many complications exist in attempts to compile estimates of global average surface-air temperature and other variables.[18] To provide a sufficiently lengthy record, resort is made to surface measurements of a wide variety of types, quality, and extent. Shortcomings exist because of changes in measurement method (e.g., canvas bucket to intake engine temperature for sea-surface temperatures), measurement locations and environment (e.g., the effects of urbanization), time of day of measurement, measurement accuracy (e.g., only to the nearest degree), varying spatial distribution of measurements, and many other factors. Despite these many difficulties with the data, better data would require upgrading the present multibillion-dollar global system for gathering weather data so that it could be more useful for monitoring the global climate. This would clearly be very expensive and require a commitment of many

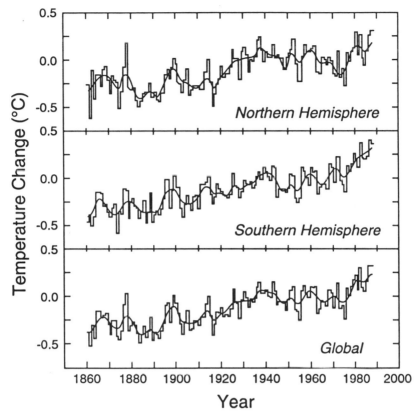

Figure 2. Area-weighted estimates of annual temperature departures from a reference normal for the Northern Hemisphere, Southern Hemisphere, and global land and ocean areas for the period since 1860. (Jones, Wigley, and Wright 1986)

years to provide a basis for examining climatic trends. At this time, we must therefore rely on what we have, despite the shortcomings.

Global compilations of available observations, which include a variety of approaches to account for possible data problems, suggest that average temperatures have increased about 0.5 ± 0.2° C since about 1850, as shown in figure 2.[19] Such a trend is seemingly in better agreement with the lower half of the NRC sensitivity range than the upper half that current model simulations appear to favor.

Separate analyses of the Northern and Southern Hemisphere records also raise questions that must be answered to refine our understanding of the quantitative relationship between atmospheric composition and climate. The Northern Hemisphere record shows a cooling in the late nineteenth century (probably due mainly to the climatic consequences of a series of major volcanic eruptions, but perhaps even a residual extension of the earlier Little Ice Age—a cool period during the seventeenth and eighteenth centuries, especially evident in land areas bordering the North Atlantic), a warming to the 1930s (particularly in high-latitude regions), a cooling into the 1970s, and a warming in the 1980s (mainly in low- and mid-latitude regions). Over rather large areas, particularly the eastern United States, cooling seems to have continued into the 1980s, keeping the average temperatures over the continental United States little changed this century.[20] The cause of this irregular hemispheric warming may simply be natural variability or may also involve other perturbing factors. Possible examples include an increase in the anthropogenic sulfate aerosol loading causing an increase in cloud albedo (particularly in industrialized regions), volcanic aerosol injections, solar variability, ocean circulation changes, switching between possibly different circulation states, a nonlinear response to the CO_2/greenhouse forcing, or other factors.

In the Southern Hemisphere the warming has been steadier and seemingly larger than in the Northern Hemisphere. Given that the extensive land areas of the Northern Hemisphere have a lower heat capacity than the oceans of the Southern Hemisphere, that the warming is steadier in the Southern Hemisphere is not a surprise, but that it is larger is somewhat perplexing.

Taken together, the model simulations and the observed warming trend suggest that the potential global warming during the next century could well be several degrees but that our confidence is not better than a factor of two.

THE CHALLENGE TO SOCIETY

The global environment is an essential resource for many if not all societal activities. Climate is an important influence on the environment, providing warmth and precipitation for agriculture and a suitable habitat for societal activities. Although the climate is not necessarily optimum in all areas, society has invested immense resources to adapt to present conditions (e.g., dams, aqueducts, buildings, etc.); thus, climatic change, especially if rapid or large, could induce significant stress, depending on the resilience and adaptability of the various activities. Even if slow, the likelihood is that small climatic changes, if continued and accumulated over a few centuries, could force significant alterations in the distributions of global population and of agricultural and forest areas.

An important complication arises because virtually all of the greenhouse gases, once injected, will remain in the atmosphere for decades to centuries. The persistence of these compounds will lead to a new equilibrium climate. The oceans moderate the rate at which the climate approaches this new equilibrium and have thus allowed only some of the warming to occur to which we have become committed by past emissions. However, the ocean delay will not prevent the new, higher equilibrium temperature from eventually being reached. In addition, climatic effects will continue because emissions of already-produced gases (e.g., chlorofluorocarbons) have not yet all occurred, being slowed in many cases by temporary containment in foams and refrigeration equipment. Nor has the atmosphere come into a new chemical equilibrium with the released gases. Thus, even drastic actions at a given time (e.g., halting all new releases) cannot prevent further changes from occurring, although such actions could limit or moderate the amount of further change. The generation, use, and, ultimately, the emission of radiatively and chemically active gases occur, in almost every case, as a consequence of seemingly beneficial and essential (at least in narrow economic terms) societal activities, including provision of food, fiber, lighting, refrigeration, insulation, home heating, transportation, and medical services. The pervasive role of these gases means that controls and alternatives must be comprehensive and global in character; as a result, changes could be costly and slowed by the extensive effort needed to find alternatives and introduce replacements. Arbitrary or abrupt changes in the availability of such benefits could significantly affect the present standard of living in many areas.

The underlying challenge is for industrialized society to achieve a balanced and sustainable coexistence with the environment, one that permits use of the environment as a resource, but in a way that preserves its vitality and richness for future generations. Meeting this challenge will require development of an approach that, although recognizing the still tentative nature of the findings, encourages the countries of the world with their varying interests and concerns to respond in a timely and coordinated way. It will not be easy to minimize projected long-term environmental and societal disruptions while retaining the potential for nations, particularly the less well developed, to continue to improve their standards of living. But the challenge to transform our ways before our world is irrevocably changed could go far toward displacing militarization and the ever-increasing push for greater national consumption as the primary driving forces behind industrial activity.

CONCLUSIONS

Model calculations, supported by paleoclimatic and analytic studies and verified against a variety of cases of past climatic change, suggest that (the global average surface-air temperatures will increase several degrees during the next century if the increasing rates of emission of greenhouse gases continue.) Such an increase would raise the long-term global average temperatures to levels not experienced in mid-latitudes back at least as far as the last interglacial period 125,000 years ago—(and the change would occur more rapidly than ecosystems have been able to adapt to such changes in the past.) Model projections of low-latitude temperature increases of a few degrees would raise temperatures to levels not experienced in tens of millions of years. (That the models indicate such high sensitivity in low latitudes, however, may be an indication that models are missing an important temperature-stabilizing mechanism that has been hidden because the observed seasonal variation in temperatures in these regions is so small.)

High-confidence predictions of global-scale temperature increases of such magnitude may provide sufficient information for the world to institute measures to slow the rate of increase of emissions and thereby the rate of temperature increase. Reducing total global emissions will be very difficult, however, without halting the increasing energy use and rising standard of living in the developing world.

Because continuation of at least present emissions levels seems highly probable, projections of potential changes at the regional level are needed to plan possible adaptive measures. Unfortunately, the reliability of the details of such forecasts is rather poor, so that decisions about whether a region must focus its response on increased winter precipitation or greater summer drying, or both, or neither, cannot yet be made. This does not mean that nothing can be done; rather it means that we must focus on increasing the flexibility and resilience of our activities that depend on the historic stability of the climate.

Thus, climate model results suggest that potential global environmental change may justify an ameliorative policy of reducing current emissions of man-made greenhouse gases but that expensive and comprehensive adaptive actions should generally await more certain results from improved models. While we labor at improving our models, we should also be identifying society's vulnerabilities to climatic change and setting in place programs to moderate potential impacts.*

ACKNOWLEDGMENT

This work was sponsored by the U.S. Department of Energy Atmospheric and Climate Research Division and performed by the Lawrence Livermore National Laboratory under Contract W-7405-ENG-48.

*A more complete discussion of the policy implications of global warming and possible and appropriate policy options for the near term is contained in MacCracken 1990.

NOTES

1. MacCracken and Luther 1985a; Schlesinger 1988.

2. See Washington and Meehl 1984; Manabe and Wetherald 1987; Hansen et al. 1984; Schlesinger and Zhao 1989; and Wilson and Mitchell 1987.

3. Schlesinger and Mitchell 1987.

4. Charney 1979; NRC 1983.

5. Cess et al. 1989.

6. Ramanathan et al. 1989.

7. Grotch 1988. W/m^2 = watts per square meter. The peak overhead solar flux is about 1370 W/m^2 at the top of the atmosphere; the global average over 24 hours at the top of the atmosphere is about 340 W/m^2, about half of which passes through the atmosphere and clouds and is absorbed at the surface.

8. COHMAP 1988.

9. E.g., Sperber et al. 1987.

10. MacCracken and Luther 1985a.

11. E.g., Budyko and Sedunov 1988; Borzenkova 1988.

12. Manabe and Stouffer 1980.

13. Grotch 1988.

14. Ramanathan et al. 1985; Wuebbles and Edmonds 1988.

15. E.g., Hansen et al. 1988; Manabe and Stouffer 1988.

16. Hansen et al. 1988.

17. Hoffert and Flannery 1985; Wigley and Schlesinger 1985; Hansen et al. 1988. An interesting corollary to the uncertainty in lag time concerns the associated thermal expansion of ocean waters and consequent sea-level rise. The long ocean lag times result when the oceans are assumed to be rapidly mixing heat from upper to lower layers; the short ocean lag times arise when deep ocean coupling is assumed to occur primarily as a result of polar bottom-water formation processes. When estimating potential sea-level rise out to the year 2100, the thermal expansion contribution to sea-level rise is about 2 to 3 times larger for the long ocean lag times for comparable climate sensitivities and emissions scenarios (e.g., 90 cm vs. 40 cm) (Frei, MacCracken, and Hoffert 1988). The sea-level rise over the past one hundred years has been 10 to 15 centimeters, presumably due both to thermal expansion and melting of mountain glaciers. Future sea-level change is also expected to result from both factors.

18. MacCracken and Luther 1985b; Ellsaesser et al. 1986.

19. Jones, Wigley, and Wright 1986; Hansen and Lebedeff 1987.

20. Karl, Baldwin, and Burgin 1988.

REFERENCES

Borzenkova, I. I. 1988. *Global climate sensitivity to the change of atmospheric gas composition from paleoclimatological data.* Leningrad, USSR: State Hydrological Institute.

Budyko, M., and Yu. S. Sedunov. 1988. *Anthropogenic climatic changes.* Report prepared for the World Congress, Climate and Development, Climatic Change and Variability and Resulting Social, Economic and Technological Implications. Hamburg, 7–10 November 1988.

Cess, R. D., et al. 1989. Intercomparison and interpretation of cloud-climate feedback as produced by thirteen atmospheric general circulation models. *Science* 245:513–516.

Charney, J. 1979. *Carbon dioxide and climate: A scientific assessment.* Report of an

Ad Hoc Study Group on Carbon Dioxide and Climate, J. Charney, chairman, National Research Council. Washington, D.C.: National Academy of Sciences.

COHMAP Members. 1988. Climatic changes of the last 18,000 years: Observations and model simulations. *Science* 241:1043–1052.

Dickinson, R. E., and R. J. Cicerone. 1986. Future global warming from atmospheric trace gases. *Nature* 319:109–115.

Ellsaesser, H. W., M. C. MacCracken, J. J. Walton, and S. L. Grotch. 1986. Global climatic trends as revealed by the recorded data. *Reviews of Geophysics and Space Physics* 24:745–792.

Frei, A., M. C. MacCracken, and M. I. Hoffert. 1988. Eustatic sea level and CO_2. *Northeastern Journal of Environmental Science* 7, no. 1:91–96.

Grotch, S. L. 1988. *Regional intercomparisons of general circulation model predictions and historical climate data.* U.S. Department of Energy Report DOE/NBB-0084. Washington, D.C.

Hansen, J., and S. J. Lebedeff. 1987. Global trends of measured surface air temperature. *J. Geophys. Research* 92:13,345–13,372.

Hansen, J., et al. 1984. Climate sensitivity: Analysis of feedback mechanisms. In *Climate processes and climate sensitivity,* ed. J. E. Hansen and T. Takahashi. Washington, D.C.: American Geophysical Union.

———. 1988. Global climate changes as forecast by Goddard Institute for Space Studies three-dimensional model. *J. Geophys. Research* 93:9341-9364.

Hoffert, M. I., and B. P. Flannery. 1985. *Model projections of the time-dependent response to increasing carbon dioxide.* U.S. Department of Energy Report DOE/ER-0237. Washington, D.C.

Jones, P. D., T. M. L. Wigley, and P. B. Wright. 1986. Global temperature variations between 1861 and 1984. *Nature* 322:430–434.

Karl, T., R. G. Baldwin, and M. G. Burgin. 1988. *Time series of regional season averages of maximum, minimum, and average temperature, and diurnal temperature range across the United States: 1901–1984.* Historical Climatology Series 4–5. Asheville, N.C.: National Oceanic and Atmospheric Administration National Climatic Data Center.

MacCracken, M. C., and F. M. Luther, eds. 1985a. *Projecting the climatic effects of increasing carbon dioxide.* U.S. Department of Energy Report DOE/ER-0237. Washington, D.C.

———. 1985b. *Detecting the climatic effects of increasing carbon dioxide.* U.S. Department of Energy Report DOE/ER-0235. Washington, D.C.

MacCracken, Michael C. (Chair). 1990. *Energy and climate change: Report of the DOE Math-Laboratory Climate Change Committee.* Chelsea, Mich.: Lewis Publishers.

Manabe, S., and R. J. Stouffer. 1980. Sensitivity of a global climate model to an increase of CO_2 concentration in the atmosphere. *J. Geophys. Research* 85:5529-5554.

———. 1988. Two stable equilibria of a coupled ocean-atmospheric model. *J. of Climate* 1:841–866.

Manabe S., and R. T. Wetherald. 1987. Large scale changes of soil wetness induced by an increase in atmospheric carbon dioxide. *J. of Atmos. Sciences* 44:1211-1235.

National Research Council (NRC). 1983. *Changing climate*. Report of the Carbon Dioxide Assessment Committee. Washington, D.C.: National Academy Press.

Ramanathan, V., R. D. Cess, E. F. Harrison, P. Minnis, B. R. Barkstrom, E. Ahmad, and D. Hartmann. 1989. Cloud-radiative forcing and climate: Results from the earth radiation budget experiment. *Science* 243:57–63.

Ramanathan, V., H. B. Singh, R. J. Cicerone, and J. T. Kiehl. 1985. Trace gas trends and their potential role in climate change. *J. Geophys. Research* 90: 5547-5566.

Schlesinger, M. E., ed. 1988. *Physically-based modeling and simulation of climate and climatic change, Parts 1 and 2*. NATO Advanced Science Institutes Series. Dordrecht: Kluwer Academic Publishers.

Schlesinger, M. E., and J. F. B. Mitchell. 1987. Climate model simulations of the equilibrium climatic response to increased carbon dioxide. *Rev. of Geophysics* 25:760–798.

Schlesinger, M. E., and Z. C. Zhao. 1989. Seasonal climatic changes induced by doubled CO_2 as simulated by the OSU atmospheric GCM/mixed-layer ocean model. *J. of Climate* 2:459–495.

Sperber, K. R., S. Hameed, W. L. Gates, and G. L. Potter. 1987. Southern oscillation simulated in a global climate model. *Nature* 329:140–142.

Washington, W. M., and G. A. Meehl. 1984. Seasonal cycle experiment on the climate sensitivity due to a doubling of CO_2 with an atmospheric general circulation model coupled to a simple mixed-layer ocean model. *J. Geophys. Research* 89:9475-9503.

Wigley, T. M. L., and M. E. Schlesinger. 1985. Analytical solution for the effect of increasing CO_2 on global mean temperature. *Nature* 315:649–652.

Wilson, C. A., and J. F. B. Mitchell. 1987. A doubled CO_2 climate sensitivity experiment with a global climate model including a simple ocean. *J. Geophys. Research* 92, no. 11:13,315–13,343.

Wuebbles, D. J., and J. Edmonds. 1988. *A primer on greenhouse gases*. U.S. Department of Energy Report DOE/NBB-0083. Washington, D.C.

THREE

Our Changing Atmosphere:
Trace Gases and the Greenhouse Effect

F. Sherwood Rowland

The radiation that reaches the earth from the sun is usually described in terminology based upon the detection capabilities of the human eye: visible radiation with wavelengths between 400 nm (violet) and 700 nm (red); ultraviolet or UV (< 400 nm); and infrared or IR (> 700 nm). The energy of the radiation increases as the wavelength gets shorter, and photodecomposition of most molecules requires visible or ultraviolet radiation. For example, the formation of ozone, O_3, depends upon the splitting of an O_2 molecule by the absorption of UV radiation, and the capture of each released O atom by another O_2 molecule. Most solar energy, however, reaches the Earth as visible radiation and, because the atmosphere is transparent to it, penetrates all the way to the surface.

This total solar energy entering the Earth's atmosphere must be balanced by an equivalent amount of energy leaving the atmosphere. While the incoming energy is mostly in the visible and UV ranges, the outgoing radiation is all in the IR because the wavelengths emitted by a body are approximately inversely proportional to the temperature of that body. The typical solar wavelength coming into the earth's atmosphere is in the yellow around 500 nanometers, after emission from a sun whose surface temperature is almost 5800° Kelvin. The Earth, in contrast, has a temperature of about 290° K, a factor of 20 lower than the surface temperature of the sun, and therefore emits radiation at wavelengths about 20 times longer than that of the incoming visible radiation. Multiplication by 20 of the 500-nanometer wavelength for yellow light gives 10,000 nanometers, or a wavelength of 10 microns in standard infrared terminology.

These IR wavelengths carry insufficient energy when absorbed by a molecule to cause its decomposition. They can, however, cause internal excitation of the molecule. Whether or not the IR radiation *will* cause that excitation depends upon whether it is absorbed, and that in turn relies upon a specific match between the IR wavelengths and energies and the possible vibrational energies for the individual molecule that is exposed. Only certain vibrational energies can be accommodated within a molecule, with other energies both larger and smaller excluded.

The three main components of the earth's atmosphere—nitrogen, oxygen, and argon—have negligible IR absorption capabilities. Monatomic argon has no infrared vibrational spectrum, and the diatomic N_2 and O_2 are quite inefficient in absorbing infrared radiation. The Earth's atmosphere, however, naturally contains several molecules that absorb IR radiation well. These include carbon dioxide, ozone, and water vapor, all of which are present as a result of natural processes. Each has three atoms, and has therefore either three or four vibrational frequencies (four for linear OCO and three each for bent H˚H and O O). Each of these molecules has a much greater capability of absorbing infrared radiation than either oxygen or nitrogen, with the energy going first into excited vibration of the molecule and ultimately into heat.

The *greenhouse effect* is caused by this interception of IR radiation leaving the Earth's surface by molecules that have more than two atoms, with CO_2, H_2O, and O_3 the primary examples in the natural atmosphere. After absorption, the energy can be reemitted, but in all directions, so that most of it does not escape into space but instead is retained within the atmosphere. Because of the necessity for incoming and outgoing energies to balance each other, the failure of escape to space by some wavelengths of IR radiation requires that more radiation go out through those wavelengths which are not intercepted. The mechanism for increasing this escape at the transparent IR wavelengths is simple—just warm up the Earth to increase the energy output at all wavelengths. The temperature of the Earth rises until enough IR radiation flows out at other wavelengths to make up for the non-escape of that trapped by CO_2, H_2O, and O_3.

In the natural atmosphere, then, some of the IR exit routes are blocked by absorption in molecules such as carbon dioxide, ozone, and water vapor, and more energy goes out through those wavelengths which the atmosphere does not absorb. These transparent regions of the atmosphere allow unimpeded IR transmission because those wavelengths do not match any of the characteristic vibrations of the important natural molecules, i.e., CO_2, H_2O, and O_3. Calculations indicate that if the Earth had an atmosphere that had only nitrogen, oxygen, and argon (of course, this is not the case), then the average temperature of the Earth would be down in the vicinity of 254° K. Instead, Earth's average temperature is about 288° K. The reason for this discrepancy is simply that the *natural* atmosphere furnishes sufficient IR absorption that the temperature of the earth has adjusted to be 30° to 40° K warmer

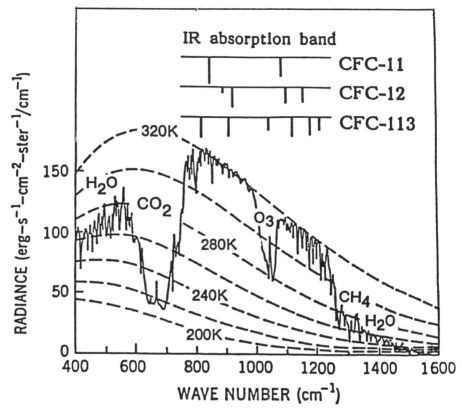

Figure 1. Infrared emissions over Sahara Desert as observed by IRIS-D instrument on Nimbus-4 satellite. Expected intensities for different temperatures are indicated by dashed lines. Emissions direct from surface (at 320° K or 117° F) are observed in "transparent" regions between 800–1000 cm⁻¹ and 1100–1250 cm⁻¹; strong absorption regions for CO_2, O_3, H_2O, and CH_4 are indicated. The main absorption positions for the three CFCs are indicated above and are mostly in the transparent regions—making incremental additions of CFCs far more efficient *per molecule* in retaining terrestrial infrared emissions than CO_2, O_3, or CH_4.

(or 54° to 72° F) to force the radiation out through the transparent wavelength windows in the atmosphere and maintain an energy balance with the incoming solar energy.

The existence of transparent and opaque bands in the atmosphere is a well-established fact, already clearly demonstrated by satellite data two decades ago. Figure 1 shows the infrared spectrum escaping from the Earth as seen by the nadir observations of the Nimbus-3 satellite over the Sahara Desert in the year 1970. The surface temperature of the desert was approximately 320° K (117° F), and the upper dashed curve shows the emission intensity expected if all of this IR were able to pass upward directly into space with no atmospheric absorption. For some wavelength regions, the Earth is essentially transparent, and the observed emissions match the 320° K curve expected for the hot desert sands. For other wavelengths, however, the emissions are greatly reduced below the 320° K line because of absorption by CO_2, O_3, and H_2O as marked. Already twenty years ago and presumably for eons before, Earth exhibited some regions of unimpeded emission, and others of heavy absorption, consistent with the 34° K temperature increase calculated as the greenhouse effect in the Earth's atmosphere. In other spectra taken from the same satellite, the Pacific Ocean at midnight exhibits in the transparent regions the surface temperature of the nighttime equatorial Pacific, approximately 300° K, and the familiar heavy absorption by carbon dioxide, ozone, and water.

The existence of this natural greenhouse effect is therefore both well understood and experimentally verified by global satellite observations. Our primary current question is this: Is mankind putting into the atmosphere enough additional molecules of CO_2, H_2O, O_3, or other IR-absorbing molecules to increase significantly the ability of the atmosphere to hold in IR radiation? If so, then the terrestrial response to lessened IR escape at those wavelengths will simply be to raise the temperature of the Earth to push a little bit more out through the transparent windows. This extra warming will force still more IR radiation out through the transparent regions—instead of a 34° warming from the greenhouse, an average of 36° or 38° may be required. Because we have experienced the adjustment of 30°–40° warming for thousands of years we think of it (correctly) as normal and speak of the greenhouse effect in 1989 in terms of the prospective increment of another 2° or 4° above the long-term base. The suggestion that the CO_2 released by the burning of carbonaceous fossil fuels such as oil, coal, and natural gas might cause such an elevation of the greenhouse temperature had already been made in the 1890s and was raised again more urgently in the 1950s and 1960s. The worldwide growth in atmospheric CO_2 was established in the 1960s through the measurements of Professor C. D. Keeling of the Scripps Institution of Oceanography. His measurements continue, as shown in figure 2.

A very important factor in the scientific evaluation of "greenhouse warming" during the last decade has been the realization that this is not just a problem of increasing CO_2 but is rather a more general problem of increasing concentrations of many trace gases. Atmospheric measurements (fig. 2) have clearly shown a CO_2 increase of about 12 percent since 1958, from 315 parts per million by volume (ppmv) in 1958 to 355 ppmv in 1989. In the mid-1970s, the chlorofluorocarbon (CFC) gases such as CCl_3F and CCl_2F_2 were shown first to be accumulating in the

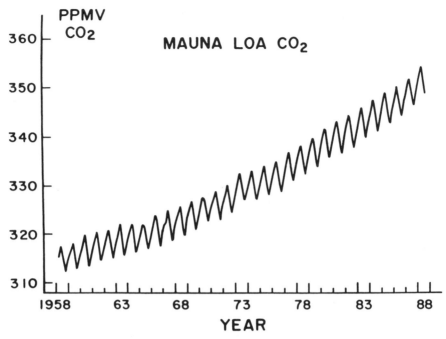

Figure 2. Carbon dioxide concentrations in parts per million by volume as measured at Mauna Loa, Hawaii, by C. D. Keeling of Scripps Institution of Oceanography.

atmosphere and then to be important as greenhouse gases. Early in the 1980s, methane (CH_4), as shown in figure 3, and nitrous oxide (N_2O) were also proven to be increasing their atmospheric concentrations.

The observed increase rates for several trace gases are given in table 1, expressed in parts per trillion (10^{12}) by volume (pptv). In these units, the carbon dioxide concentration in the 1989 atmosphere is about 355,000,000 pptv, and the yearly rate of increase is about 1,500,000 pptv. This absolute rate of increase for CO_2 in pptv/year is very much larger than for any other atmospheric trace gas. For example, the yearly *increase* in carbon dioxide is more than a thousand times larger than the *total* concentration of the chlorofluorocarbons in the atmosphere. Chlorofluorocarbon-12 (CCl_2F_2) is increasing only by 16 or 17 pptv per year with a concentration around 450 pptv in the late 1980s.

How it is that one molecule (CCl_2F_2) increasing at 17 pptv per year can possibly play a significant role in the greenhouse effect, when another (CO_2) is going up by 1,500,000 pptv per year, almost 100,000 times faster?

The structure of the CFC molecules is such that their individual molecular vibrations furnish very strong IR absorption at different wave-

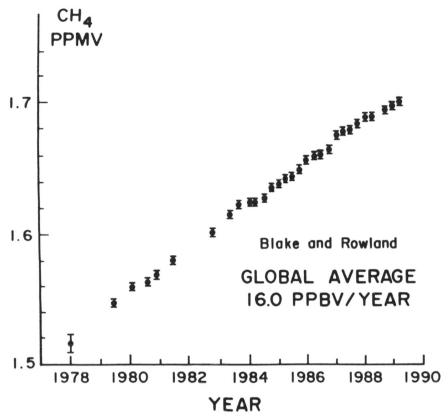

Figure 3. Globally averaged concentrations of methane, as measured in the Pacific Region between 71°N-47°S latitudes by D. R. Blake and F. S. Rowland.

lengths from CO_2, H_2O, and O_3 and fall right in the middle of the transparent wavelength regions, as indicated in figure 1. An additional molecule absorbing in the transparent region is very much more efficient in retaining IR radiation than one additional molecule of CO_2, because the latter can only absorb the same wavelengths being swallowed up by the 355,000 pptv of CO_2 already present in the atmosphere. In an artificial atmosphere free of both CO_2 and CFCs, individual molecules of CO_2 and CFC are roughly comparable in retaining IR radiation. In Earth's relatively CO_2-rich atmosphere, however, the addition of more CO_2 is a form of overkill in IR absorption and is not very effective *per added molecule*, as shown in table 1. Nevertheless, the sheer magnitude of the yearly 1,500,000 pptv increase in CO_2 is enough to retain for carbon dioxide increases the most important role in the greenhouse effect.

As shown in table 1, the chlorofluorocarbons are about 20,000 to 25,000 times as efficient as added carbon dioxide in the 1989 atmo-

TABLE 1. Contributions to the Greenhouse Effect
from Concentration Changes in Atmospheric Trace Gases

Gas	Relative 1989 Concentration (pptv)	Yearly Change in Concentration (pptv/year)	Relative Greenhouse IR Efficiency (from D. Wuebbles)	Relative Incremental Greenhouse Contribution, 1989 (Yearly Change × Efficiency)
CO_2	355,000,000	1,500,000	1	1,500,000
CH_4	1,700,000	16,000	31	500,000
N_2O	300,000	600	200	120,000
CCl_2F_2 (CFC-12)	450	17	25,000	425,000
CCl_3F (CFC-11)	270	11	22,000	240,000
CCl_2FCCl_2F (CFC-113)	90	7	25,000 est.	175,000
Total for 5 gases, excluding CO_2				1,460,000

sphere. Methane is about 30 times as efficient as CO_2 in the Earth's atmosphere. These high relative efficiencies, taken with the known concentration increases, cumulatively have greenhouse impacts that begin to match that of carbon dioxide.

How sure are scientists that these trace gases are increasing in average global concentration? This is a question for which some of the answers have been provided by the trace gas measurements performed by my own research group. Our involvement in atmospheric chemistry started with the observations in 1971 by James Lovelock that the molecule trichlorofluoromethane (CCl_3F) was present at a level of about 60 or 80 pptv in the north temperate zone and about 40 pptv south of the equator. Contemplation of these findings led to scientific curiosity and eventually the questions: What will happen to the CFC molecules in the atmosphere? How long will it take?

The answer to the first question is now often called the Rowland-Molina theory of stratospheric ozone depletion by chlorofluorocarbons. The CFCs eventually are blown apart by solar UV in the mid–stratosphere, releasing atomic Cl, which then proceeds to destroy globally significant amounts of ozone by a chain reaction process. The full scientific description of ozone depletion is a separate topic and has been discussed extensively elsewhere.

The important aspect of the Rowland-Molina theory as far as the greenhouse effect is concerned is contained in the answer to the "how long" question: the CFCs last many decades. The time scale for the removal of CFC-11 (CCl_3F) from the atmosphere is about 75 years on the average, while its companion molecule, CFC-12, requires approximately

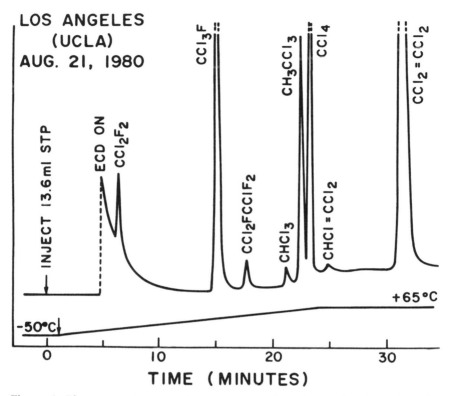

Figure 4. Electron capture gas chromatogram of an air sample collected on the UCLA campus on August 21, 1980—a typical urban sample. The time on the horizontal axis serves to identify the individual trace gases present in the sample, and the height on the vertical axis measures the quantity present. Because the individual molecular sensitivities are different, the peak heights are separately calibrated. The vertical dashed lines indicate off-scale peaks.

120 to 140 years. While such long lifetimes were originally estimated theoretically by Rowland and Molina, they have now been confirmed by actual measurements in the atmosphere itself. From the differences between the total amounts released and still present in the atmosphere, now measured over a time period of more than 15 years, the atmospheric lifetimes have been established to be fully as long as originally calculated.

Figure 4 shows a "gas chromatogram" of air taken from the UCLA campus in Los Angeles in 1980. The chemical identity of different compounds is established by the distance along the horizontal time axis: each molecule has its own characteristic travel time through the sixty-meter-long gas chromatographic column. The atmospheric concentration of each compound is shown by the vertical height of its corresponding

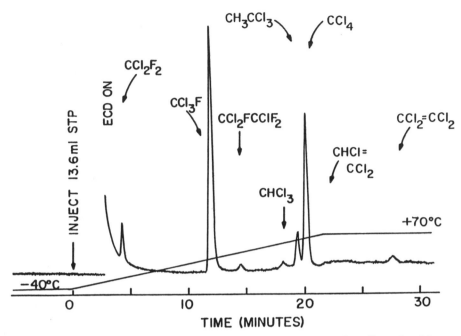

Figure 5. Electron capture gas chromatogram of an air sample collected at No-sappu Point on the Japanese island of Hokkaido on August 3, 1979—a typical remote-location sample. The vertical scale is the same as in figure 4.

peak. (Because the instrument sensitivity varies widely with different compounds, each peak height requires its own calibrated scale for conversion into pptv units.) Three successive CFC peaks—CFC-12, CFC-11, CFC-113 (CCl_2FCClF_2)—are followed by chloroform ($CHCl_3$), methyl chloroform (CH_3CCl_3), carbon tetrachloride (CCl_4), trichloroethylene ($CHCl = CCl_2$), and tetrachloroethylene ($CCl_2 = CCl_2$). These peaks are quite representative of the usual composition of air in urban environments—the standard loading found in cities into whose atmosphere a lot of chlorinated molecules are being released. A similar 1980 air sample taken in San Jose, California, showed the same eight molecules, albeit with much more CFC-113, because this molecule was already then widely used for cleaning electronics. The large CFC-113 peak was a characteristic of San Jose air—a trademark for Silicon Valley and its electronic industries—and is even more prominent in 1989.

However, the global concentration of these molecules cannot be estimated from such data taken in the cities, which are the major sources for the man-made CFCs. Rather, extensive sets of observations must be made in remote locations. Figure 5 shows the concentration data from the easternmost tip of the island of Hokkaido in Japan, a very remote location. Seven of the same eight compounds are present, but in much smaller concentrations. The concentration of $CHCl = CCl_2$ is almost undetectable, because it has a survival time in the atmosphere of only a few weeks, and the Hokkaido air had traveled for many weeks without passing over a major city.

For eleven years I and the members of my research group have been traveling regularly to remote locations from Alaska to the South Pole, bringing back air samples for trace gas analysis. Our data and those of several other research groups have clearly shown rapid worldwide increases in the concentrations of each of the CFCs. The growth in the organochlorine concentrations in the atmosphere is graphed in figure 6, with actual data back to 1971. Extrapolations farther backward in time are based on known production figures, while forward estimates are based on various hypotheses for future global emission trends.

Methyl chloride, CH_3Cl, is the only molecule shown in figure 6 for which an important known natural source exists. (Kelp beds produce the related compound methyl iodide, CH_3I. As methyl iodide percolates up through seawater, chloride ion displaces the iodine, forming CH_3Cl.) As far as we can determine, the 600 pptv concentration of CH_3Cl is almost entirely natural, and the measured values have not changed during more than a decade of observations. In contrast, CCl_4 , CFC-11, CFC-12, CFC-113, and CH_3CCl_3 have been steadily adding to the total chlorine loading of the atmosphere. Whereas the atmosphere is assumed to have contained about 0.6 parts per billion by volume (ppbv) in 1900 (essentially only CH_3Cl), the total chlorine loading was about 0.8 ppbv in 1950, 1 ppbv about 1965, 2 ppbv around 1976, and is now approximately 3.5 ppbv in 1989. Worldwide, there has been a very substantial increase in the total amount of atmospheric chlorine, and most of this increase has been accounted for by the growth of these long-lived CFC compounds.

Each of these three CFC gases absorbs strongly in the transparent region of the infrared. As shown in table 1, the CFCs together make an atmospheric greenhouse contribution that significantly adds to that of carbon dioxide. If we were to continue putting these molecules into the atmosphere at the rate we were doing in 1986, then the concentrations will rise as shown by the thin solid lines in figure 6, with a total atmospheric chlorine concentration reaching 5 ppbv early in the next century. The provisions of the United Nations Montreal Protocol, agreed to by all of the major nations in September 1987, took effect in July 1989 and call initially for a cutback in CFC emissions to the 1986 levels. The Montreal Protocol then requires a 20 percent cutback from the 1986 levels in 1994, and a 30 percent additional cutback in 1999. These reductions would produce the chlorine concentrations shown by the dotted lines (fig. 6). The Protocol as now established does not really make

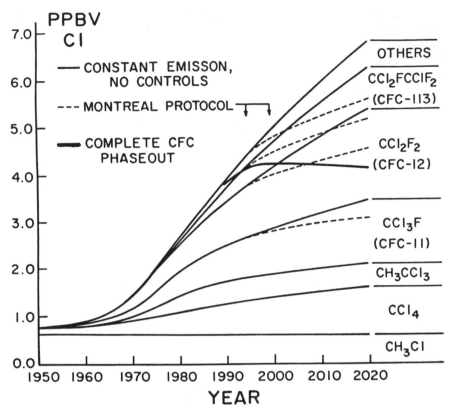

Figure 6. Averaged Northern Hemispheric chlorocarbon concentrations in the atmosphere at surface level. Actual measurements from early 1970s to the present. Extrapolated back to 1950 from known production figures. Extrapolated forward to 2020 with several different assumptions: (a) continued yearly emissions at 1986 levels (thin solid lines); (b) yearly emissions modified according to Montreal Protocol of September 1987—reduction to 80 percent of 1986 levels in 1994, and to 50 percent of 1986 levels (dashed lines); and (c) complete phaseout of CFC emissions by the year 2000 (thick solid line).

much difference until the next century, and total chlorine would still rise rapidly, although not as rapidly as without it.

The terms of the Montreal Protocol are in the process of being greatly strengthened. For instance, the European Economic Community announced in March 1989 that its twelve countries will no longer be emitting chlorofluorocarbons by the end of the century. The major U.S. manufacturers of CFCs such as DuPont and Allied-Signal have also said that they will phase out the production of chlorofluorocarbons by the end of the century. If these actions are carried through as planned, much of the problem of the growth in CFCs will be taken care of within the next decade or so. The growth of total atmospheric chlorine concentration should peak between 4.0 and 4.5 ppbv at the end of the decade, and then gradually decrease. Even with the total phaseout of CFC production, however, the chlorine concentrations will remain high for all of the twenty-first century. The slow decrease as shown by the thick solid line in figure 6 is the straightforward result of the long atmospheric lifetimes of the CFCs.

In 1978 we made our initial measurements of the concentrations of methane in the atmosphere. At that time the concentration in the Northern Hemisphere was about 1.58 ppmv and in the Southern Hemisphere about 1.46 ppmv, for a global average of 1.52 ppmv. Our repeated measurements since then have shown a steady, parallel increase in both hemispheres to a 1989 global average of 1.70 ppmv, as shown in figure 3. We now make these measurements every three months, going everywhere from Alaska to New Zealand, with visits to the South Pole in the summer, bringing back air samples from which we derive a global average.

Our data show an average increase in methane of 16 ppbv/year over the time period of eleven years, a 12 percent increase. The atmosphere is certainly increasing its concentration of methane, bringing with it the ability to absorb infrared radiation in wavelengths that are different from those of carbon dioxide. The CH_4 contribution to greenhouse warming also adds significantly to that from CO_2, and, unlike the CFCs, controls will be much harder to apply.

The sources of methane in the atmosphere are roughly 75 percent biological—from microbial action in swamps, rice paddies, cattle stomachs, and so on—and perhaps 25 percent from methane released in the production and use of fossil fuels. We know rather accurately that the atmospheric lifetime of methane is about ten years, which means that 10 percent of the 4,800 megatons in the atmosphere must be replaced each year. While we know that about 500 megatons of CH_4 is released each year, assessment of the distribution among those contributors is more difficult. However, it is possible to distinguish biological methane from fossil-fuel methane, because biological methane has carbon-14 in it and fossil-fuel methane does not. Carbon-14 atoms are formed continually in the atmosphere by nuclear reactions induced by high-energy cosmic radiation and become incorporated first into $^{14}CO_2$ and then by photosynthesis into plants and ultimately into all living species. Because ^{14}C is radioactive with a half-life of 5,700 years, the carbon residues of biological species remain radioactive for periods of tens of thousands of years. Coal, oil, and natural gas, however, have been sequestered from cosmic radiation for millions of years and no longer have any ^{14}C. The measurements of the ^{14}C content of atmospheric methane lead to the interpretation that about 75 percent is biological and the rest of it probably originates from fossil fuel. (These measurements are made less precise by the "pulse" of ^{14}C introduced into the Earth's atmosphere by the testing of nuclear weapons in the atmosphere during the 1950s and early 1960s.)

Measurements of the atmospheric concentrations of CO_2 and CH_4 can now be carried out for thousands of years into the past. The major route to this information about earlier atmospheric composition is from ice cores taken out of the accumulated layers of ice lying over a bedrock base in Greenland and Antarctica. At the Vostok station operated in Antarctica by the Soviet Union, the buildings lie about 4,000 meters above the solid bedrock with intervening layers of 4,000 meters of ice gathered year by year above the rock. Vostok is necessarily at high altitude and is very cold (an *average* temperature of about $-59°$ F), with only a little snowfall each year. The Vostok ice sheets have been cored down 2,200 meters, which corresponds to an ice accumulation for 160,000 years, with little bubbles of air trapped all along the way. The present warm period (the Holocene) began about 20,000 years ago, preceded by an icy period of about 100,000 years, then another warm or interglacial period, and then another time of ice. A core going back 160,000 years extends through the most recent series of ice ages, beyond the last interglacial period, and into another earlier sequence of ice ages. Swiss and French scientists have measured the concentrations of CO_2 and CH_4 throughout this long period.

The concentration of CH_4 was no larger than 0.7 ppmv during our present warm period until the last 200 years or so, and was also about 0.6–0.7 ppmv during the last interglacial 130,000 years ago. In between, during the fluctuations of most recent ice ages, the CH_4 concentrations varied from 0.3 to 0.4 ppmv; the level was also 0.3 to 0.4 ppmv in the ice age 150,000 years ago. Over all the past 160,000 years, except the last 200, the CH_4 concentration varied from 0.3 to 0.7 ppmv through ice ages and interglacials. Changes of 0.3 ppmv in CH_4 concentration took place over millennia, varying between 0.3 and 0.7 ppmv. By comparison, when we started our measurements in 1978, the concentration was 1.52 ppmv and now, in 1989, 1.70 ppmv, and increasing 0.1 ppmv every 6 years. We are truly moving rapidly into previously uncharted areas.

The most famous series of trace gas measurements is shown in figure 2, with the CO_2 measurement made on top of Mauna Loa, Hawaii, by C. D. Keeling. The pattern shows the annual "breathing" of the Northern Hemisphere, with rapid CO_2 decreases as it is used in photosynthesis by the green plants in spring and summer together with the year-round release of CO_2 from the decay of biological material. This annual cycle is superimposed on a steadily rising background, so that the yearly average CO_2 concentration has increased about 12 percent from 315 to 355 ppmv in 31 years.

In summary, the CFCs are increasing at 5 percent per year with CFC-113 going up at a more rapid rate; methane approximately 1 percent per year; CO_2 by 0.5 percent per year; N_2O about 0.2 percent per year. These rates of increase have been fed into detailed models of the infrared absorbing characteristics of the atmosphere, and have provided the estimated relative contributions from the various trace gases as summarized in table 1.

Most such calculations were carried out before the intense concerns in the past two years over ozone depletion by CFCs. The future estimates of the CFC influences will no longer show the incremental IR contributions characteristic of the 1980s. By the year 2000, CFC emissions are likely to be nearly zero, except for residual release from existing equipment. This part of the greenhouse problem has apparently been solved on an international basis, although the rapidity of implementation of the solution over the next decade will still have important atmospheric consequences. The concern over stratospheric ozone depletion by the CFCs stopped the exponential growth in CFC production in the mid-1970s, converting it into more or less constant release for a decade before growth began in the mid-1980s. Without the restrictions and doubts placed upon CFC growth in the 1970s, the greenhouse problem in the 1990s would have been a CFC problem augmented by CO_2, rather than a CO_2 problem augmented by CFCs and other trace gases. However, nitrous oxide and methane emissions have not been curtailed or even well understood, and their concentrations will continue to grow.

Carbon dioxide is still the major contributor to the greenhouse effect, and its yearly contribution appears to be increasing. Figure 2 shows that the amount of carbon dioxide over Hawaii begins to decrease in March or April because of the seasonal growing cycle in the Northern Hemisphere. Throughout the year the continuing decomposition of plants releases CO_2, and the burning of fossil fuels releases CO_2, and these increases are not countered by photosynthesis during the late autumn and winter. Each year the spring value is a little bit higher than the year before. The magnitude of the seasonal cycle is slowly increasing from about 7 ppmv in the 1960s to 8 ppmv in the 1980s. The change is not large, but the breathing cycle is expanding.

However, after allowing for the seasonal cycle, the growth from 315 ppmv to around 355 ppmv has been very steady, although not uniform. In the 1960s the yearly increase was about 0.8 ppmv/year; in the 1980s, about 1.5 ppmv/year. In the last two or three years, the increase has been more than 2 ppmv/year. The implication is strong that the growth in CO_2 underlying the greenhouse effect is accelerating.

An important question for dealing with the greenhouse effect will be the full understanding of these CO_2 concentration changes. The total amount of carbon from the burning of fossil fuel that is going into the

atmosphere is considerably larger than the carbon dioxide increase reg-
istered in the atmosphere. This discrepancy is described by the concept
of an "airborne fraction," i.e., the fraction of CO_2 that hasn't disap-
peared from the atmosphere by going into the ocean or some other
place. The airborne fraction was about 50 percent in the 1960s but is
now approximately 60 percent. Appreciable CO_2 contributions are also
being received from the burning of the tropical forests. It is important
that we understand all the characteristics of this curve in order to understand how to predict and control future carbon dioxide concentrations in the atmosphere. Similarly, it will be important to understand the
individual contributions from various processes toward methane and nitrous oxide emissions.

The procedures necessary to solve the chlorofluorocarbon problem
have been put into place on an international scale and have begun to be
implemented. We still have left for the future, however, efforts to reduce
emissions of carbon dioxide, methane, and nitrous oxide. The advan-
tages of prevention over cure in the realm of atmospheric problems are
illustrated by the time scale for atmospheric recovery from the CFC releases of the past two decades. Stratospheric ozone depletion and infrared greenhouse absorption will be caused by the chlorofluorocarbons in
the atmosphere for all of the twenty-first century, because the lifetime
of these molecules is very long. A return to the stratospheric chlorine
concentrations of the 1960s will not happen before the twenty-third
century.

APPENDIX

1. Question: How much change has there been in water vapor in the
   atmosphere?

   Answer: The measurements on a global basis do not show any increase at this point, but such global measurements are extremely difficult because of the wide variability in atmospheric humidity with
   latitude and season. One might expect a slow increase: if the average
   global temperature goes up 0.5° C, more evaporation should occur,
   and more water vapor will go into the atmosphere. With more water
   going in, more water must eventually come out, and overall we would
   expect more rainfall on a global basis. However, the greenhouse contributions are determined by the amount of water vapor in the atmosphere, and that depends not only on the amount of evaporation
   but upon the average time until rainfall. The question is, I think, very
   open as to whether or not the atmospheric water vapor concentration
   has changed over the last twenty years or so. That is a much more
   difficult measurement than for the long-lived molecules such as CO_2,
   CH_4, and the CFCs.

For chlorofluorocarbons with lifetimes of a century or so, we see practically the same concentrations everywhere in the Southern Hemisphere when a measurement is made. As far as methane is concerned, the range in concentrations is not very great for remote locations anywhere in the world, a few percent. With water vapor, in contrast, as we know, the humidity changes from day to day. Thus it is very much harder to do an accurate global total, and therefore to detect any global changes.

2. Question: Is the possible source of some of the methane increases an indication of warmer temperatures?

Answer: The explanation for why methane is increasing can have several possible components. The control of methane and of carbon monoxide in the atmosphere is accomplished almost exclusively by reaction with hydroxyl radical. At the same time, the major removal process for hydroxyl radical is reaction with carbon monoxide and with methane. If those two statements were complete by themselves, then we would say that there is a limited, fixed capacity of the atmosphere to oxidize carbon monoxide and methane, and that we just might be exceeding that capacity.

The question of whether or not those statements are simultaneously true depends upon whether one regenerates the hydroxyl radical after it has reacted with CO or with methane. The indications are that the regeneration of hydroxyl depends upon the nitric oxide concentration of the atmosphere. In urban areas with high concentrations of hydrocarbons and of nitrogen oxides, then the radicals are regenerated. But in remote areas with low nitrogen oxide concentrations, it seems likely that the reaction of HO with CO really eliminates both. So, to some extent, the increase in methane is probably coming about because of a gradual slowing down in the ability of the atmosphere to remove these species, slowing down because there are just too many of them there to be handled. A possible part of the methane increase of 1 percent per year may be this slowing down of oxidation capability.

In addition, the chief biological sources are cattle and rice paddies and swamps. The number of cattle in the world went up about 55 percent between 1950 and 1980, from 0.8 billion to 1.3 billion. The acreage devoted to rice paddies has not changed much, but the number of crops per year has increased. The fraction of the time that the rice paddies are under water—the time when methane is given off by the rice plants—has increased. It is quite possible that the reason for some of this methane increase is simply that more methane is being put into the atmosphere.

Trying to understand all of these details requires a very thorough study. For instance, the NASA tropospheric experiments called the

Alaskan Boundary Layer Experiments have been looking at the question of gas evolution from the Alaskan tundra, and temperature is one of the variables being monitored. The ABLE-3 program in Alaska during July and August 1988 was quite extensive, with an experimental field on the ground and aircraft flyovers investigating the evolution of these gases. There was much concern regarding whether an increase in temperature of the Alaskan tundra might cause an increase in methane evolution.

Graduate student Ed Mayer and I did a study on a swamp in upper New York State ten years ago and found that methane evolution there was very much temperature-dependent. When the temperature went up, much more methane came out, but that was a very small swamp and a one-year study. Overall there is not very much information, but more and more broad programs are coming in to look at the relationships in ecosystems such as the Alaskan tundra. The same group of scientists supported by NASA are going to resume the ABLE experiments in eastern Canada (near Hudson's Bay) during the summer of 1990. The emphasis again will be placed on the gas evolution from Arctic tundra. Therefore, there are some experiments in progress, but there is really not a whole lot of information with which to unravel the individual components.

3. Question: You mentioned your Mauna Loa study—which I gather contains the longest continuous measurement since 1959. There are active volcanoes in Hawaii. Couldn't a lot of the trace gases be due to volcanic activity? Also, could you comment on the time period of the 1970s and early 1980s where there was very unusual and heavy volcanic activity, including Mt. St. Helens? What might these volcanic eruptions have to do with massive emissions of trace gases circulating in the global atmosphere?

Answer: First, let me point out that the measurements at Mauna Loa are not our measurements. These measurements were instigated by Roger Revelle for the International Geophysical Year of 1957–58, and have been carried out for the entire thirty-year period by Dave Keeling at the Scripps Institution of Oceanography. The influence of volcanoes on those measurements is zero. The high-altitude site on top of Mauna Loa occasionally gets upswept air coming up from the vegetated areas at lower altitudes. At ground level in the middle of a forest, the amount of carbon dioxide varies quite widely during the day, because photosynthesis is going on and carbon dioxide gets used up during daylight hours. This phenomenon dictates a measurement site that is well away from vegetation, and resulted in the original placement of the station at high altitude on Mauna Loa. Even then, the vegetation is more of a problem than volcanoes at Mauna Loa. The continuous daily records are carefully examined by Keeling to edit out periods of influence from the lower-altitude vegetation. If there were volcanic explosions that put anything into the atmosphere nearby, the CO_2 data would have also been examined to remove such effects as well.

In fact, CO_2 measurements are being made not just at Mauna Loa but at Barrow, Alaska, and a number of other sites. Similar CO_2 patterns are seen everywhere. In the Southern Hemisphere the annual variation is much less, but the year-to-year increase has moved along parallel to what is seen in the Northern Hemisphere. I believe that the local perturbations of the Mauna Loa record have been quite satisfactorily edited out of the data that is graphed in figure 2. As I say, similar measurements have been made in numerous places around the world, with essentially all of the others not situated near active volcanoes. Volcanic emission of CO_2 does not seem to be a serious problem; that is, the amount of carbon dioxide that is in the atmosphere and the amount that is increasing each year is a very large amount relative to the amount of CO_2 that would be contained in volcanoes. There is more concern about the particulate level in the atmosphere from volcanoes, and about the sulfur dioxide level, and one certainly sees those effects. But volcanoes certainly do not affect the measurements of carbon dioxide, methane, or the chlorofluorocarbons.

4. Question: In terms of the research going on right now in the areas that you talked about, what are the biggest areas that need more scientific attention?

Answer: The number of measurements of trace gases generally—not with satellites, but on the ground level—are on a global basis really very small. When I show you a slide that gives you a global average concentration for methane for a particular date, you are seeing results based on about sixty to seventy-five samples collected over several weeks. That is really not very many samples to represent the entire atmosphere. The only reason that we get global average concentrations with enough precision to establish trends over time is the thorough atmospheric mixing as exemplified by nearly constant values of CH_4 over the whole Southern Hemisphere.

A number of research programs are making such measurements, and I think that they can easily be intensified. The measurements need to be extended to a lot of other gases, to the other hydrocarbons. Methane has a relatively long lifetime and is present at a much higher concentration. It is therefore easier to measure than molecules such as ethane and propane. However, I think that we also need global surveys on those nonmethane hydrocarbons just to be sure that when we say we are summing up all of the trace gas effects, we really have included all the important trace gases.

FOUR

The Quest for Reliable Regional Scenarios of Climate Change

W. Lawrence Gates

It is fair to say that the demand for climate scenarios for impact estimation far exceeds the supply of reliable information that climate modelers have in their possession. Climate models certainly have systematic errors; most of these errors are poorly understood and are, in general, poorly documented. Nevertheless, the demand for scenarios and the need to use climate models in applications continue to accelerate. Modelers can only hope that this acceleration of demand and interest will somehow lead to an acceleration in modeling research, and the allocation of the resources that are needed for such work.

Climate models are a derivative of the models that are used to predict the daily weather, the so-called numerical weather-prediction models. Climate models differ from weather-prediction models in that climate researchers do not have to assume any particular initial state such as today's weather in order to go forward, and their models are integrated for much longer periods of time. The evolution of the atmosphere's three-dimensional distribution of variables such as wind, temperature, and pressure are predicted in a climate model, along with the occurrence of convection, clouds, and precipitation, and the structure of the surface boundary layer. Although they predict synoptic events as do weather models, climate models are used to derive the long-term average statistics which we regard as the model climate. Such models offer our only hope of achieving a quantitative understanding of possible future climates; the fields displayed by a climate model are physically consistent, have no missing data (in the sense that the horizontal and vertical distribution of data on the model's resolution are complete), and are objective (in the sense that they depend only on the assumptions or conditions under which the simulation is made). We begin by illustrating current climate-model performance as a prelude to the consideration of the local climate changes related to impact estimation.

CLIMATE-MODEL PERFORMANCE

Figure 1 is an illustration of a typical climate model's performance, in this case a simulation of the February surface-air temperature; this happens to be a model from the Geophysical Fluid Dynamics Laboratory (GFDL) in Princeton, which is widely considered the premier modeling group in the United States. In comparison with the distribution of the observed surface-air temperature, the model's simulation may be regarded as reasonably successful on the large scale. There are, however, errors of the model (versus observation) of the order of 5° C. Such errors, of course, are of great interest to modelers but in general have not been thoroughly diagnosed. In the case of surface-air temperature, the errors are related to the model's inadequate treatment of the exchange of heat and moisture at the surface, which is in turn related to the model's failure properly to simulate the planetary boundary layer and the associated low-level cloudiness.

Model simulations and their errors can be examined for a wide variety of variables. Figure 2 shows the average January precipitation versus latitude for a number of early climate models. There is a lot of scatter amongst these models, as well as substantial errors with respect to the observed distribution shown by the thin full line. Although many of these models have since been improved, this kind of scatter is still characteristic of climate models. In general, the models tend to overestimate precipitation, although the reasons why this is so are not yet well understood. Because it occurs over short spatial scales, precipitation and the associated distributions of convection and cloudiness are notoriously difficult to represent properly in a climate model; cloudiness and precipitation are parameterized in a climate model as sub-grid scale processes that are not directly addressed or resolved. It is ironic to note that it is just these local or small-scale phenomena with which we are often most concerned in the estimation of the impacts of climate change.

CLIMATE-MODELING STRATEGY

The methodology of climate modeling may be summarized as in figure 3. The climate or long-term averages obtained when a model is integrated under a standard set of conditions is what we refer to as the "control" climate. A model's results can be compared to actual observation, which thereby validate or verify the model. A comparison region by region and variable by variable allows the construction of a matrix of

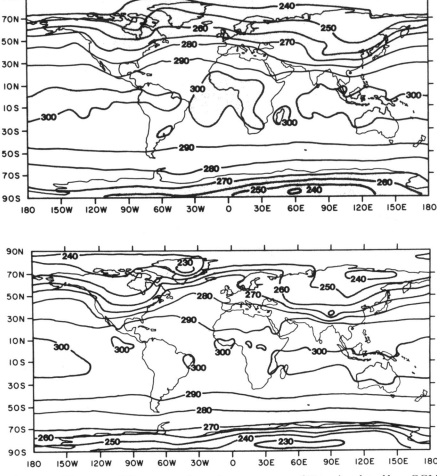

Figure 1. The mean February surface-air temperature (°K) as simulated by a GCM at GFDL (above) and as observed (below). (From Manabe and Stouffer 1980)

model errors. The aim of the climate modeler, of course, is to reduce these errors in a systematic way.

Once a control climate is validated, the model may be applied in experiments in which one or another condition is changed; an increase of CO_2 is one of the more popular experiments and is the one with which we are most concerned here. When an experiment is made in which CO_2 is doubled, the modeler "subtracts" the control climate from the experiment's climate, which then yields the modeled climate change. This is illustrated by the carbon dioxide change experiments given in figure 4, showing two independent models in which the CO_2 has either been qua-

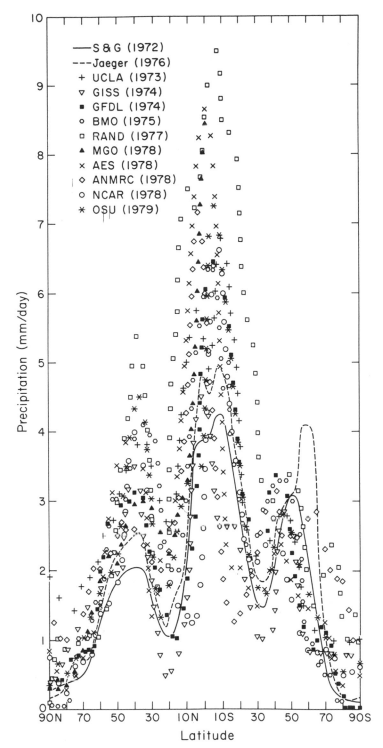

Figure 2. The mean zonally-averaged January precipitation as simulated by a variety of GCMs and two estimates of the observed climatological distribution (thin full and dashed lines). (From Gates 1987)

CLIMATE MODEL TESTING AND APPLICATION

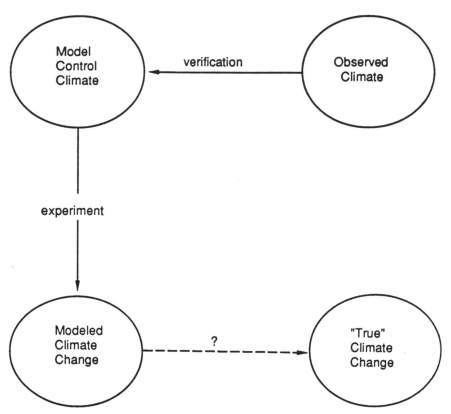

Figure 3. Elements of a climate-modeling strategy, in which the "true" climate change to be expected from experimental changes is inferred from a model's experiment (relative to a control or standard) and the model's verification errors (relative to observed climate).

drupled (upper panel) or doubled (lower panel) in reference to the models' control. Neither the patterns nor the magnitudes of the modeled changes of January average surface-air temperature are similar. Sorting this out and pinpointing which model features are responsible for these differences is a continuing and important research problem.

THE ESTIMATION OF REGIONAL CHANGES

Even though a particular climate model's results may be satisfactory on the large scale, there remains the problem of determining changes on the local or model sub-grid scale, which is everything below several hun-

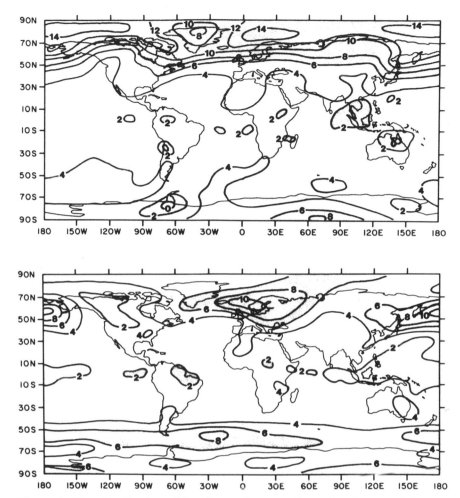

Figure 4. The distribution of the change in mean December-January-February surface-air temperature ($^\circ$ C) as a result of quadrupled CO_2 with a GCM at GFDL (above) and as a result of doubled CO_2 with a GCM at NCAR (below). (From Washington and Meehl 1984)

dred kilometers. A strategy for accomplishing this is shown in figure 5, in which the ordinate is useful climate information. In their raw output, climate models certainly have useful information on the large-scale end of this diagram. The first step in using these results is to determine the statistical significance of the modeled climate changes from suitably long control and experimental runs, a step that is not always taken in a systematic way. Assuming this has been done and that at least some of the results have been judged statistically significant, there remains the problem of going from large to small scales.

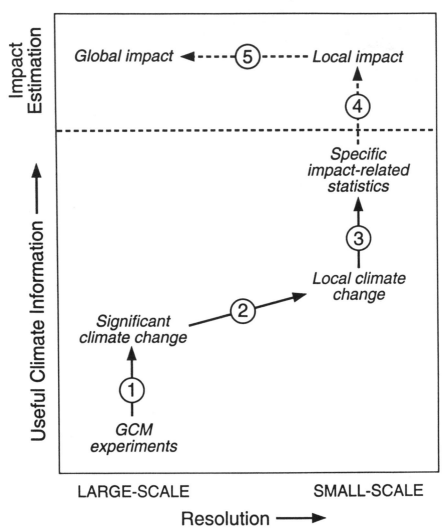

Figure 5. Elements of a strategy for the application of the results of GCM experiments to the estimation of the local impacts of climate change. See text for description of steps 1–5. (From Gates 1985)

The local climate could be estimated by interpolation of the model's solutions. Although there is no theoretical basis for such a procedure (since the solution between model grid-points is not determined), it has often been done. Taking the state of Oregon as an example of an area representing a grid point in a typical climate model, figure 6 illustrates the local geographic information. What has been done here is to take the local or station monthly averaged observation of surface tempera-

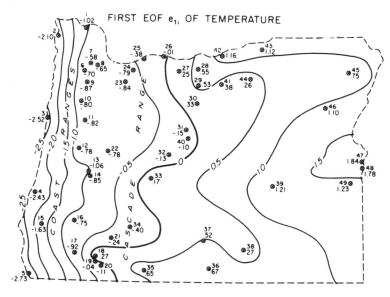

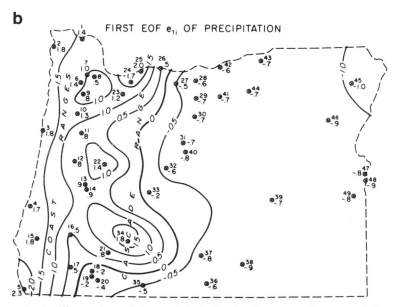

Figure 6. An illustration of the statistical distribution (over the state of Oregon) of the change of monthly mean surface-air temperature (above) and precipitation (below) that may be statistically inferred from a change of the area averages alone. (From Kim et al. 1984)

ture and precipitation from thirty years of record and ask the question, How much of the local variation from month to month could be accounted for on the basis of only the statewide area average changes? The answer is shown in terms of the contours of the covariation between the local variability and the variation of the grid-size area average. (Figure 6 actually shows the amplitude of the first eigenfunction of the correlation.) Evidently one can infer an appreciable amount of the local variability of temperature (and similarly for precipitation) from the change of the area mean alone, together with knowledge of the local climatology. By the use of such an empirical transfer function, one could estimate the local allocation of a modeled climate change such as that due to increased CO_2, assuming local climatological observations were available. Such functions clearly show the dominant local influence of features such as mountains and large water bodies.

Returning to figure 5, the approach to local scales could also be achieved by embedding a higher-resolution model within a large-scale climate model, say over the western United States or over California. This is routinely done in weather prediction, but whether or not such a technique will be successful for climate is still an open question. The success of this technique will depend critically upon our knowledge of the resolution-dependence of climate models, a matter to which relatively little attention has been given.

Once estimates of local climate change have been made, the next step is the construction of specific measures or statistics related to the local climate impacts of interest, such as those related to energy, agriculture, water use, or local ecosystems, after which the local impacts may be determined with an impact model. Finally, the large-scale impacts may be constructed by appropriate aggregation.

CONCLUSION

A number of problems still confront climate modelers if the challenge of providing information that can be used for local impact estimation is to be met. First, the models must be improved. Models continue to show large systematic errors, and the structure and behavior of these errors are not well understood. This suggests the need for more analysis, diagnosis, and intercomparison of model results, in order to understand the reasons for the differences among models and their sensitivity to both parameterization and resolution. It is to be hoped that the resources necessary to do this on a sustained and coordinated basis will be made available.

It should also be recognized that modeled climate changes will inevitably be in terms of frequency distributions rather than categorical results. Ideally, these distributions should be constructed from the statistics of an ensemble of model runs, rather than by guessing or by uncertain analogs. Such information would permit the determination of the risk or uncertainty of the derived climate impact estimates and would place climate model applications on a firmer scientific basis.

APPENDIX

1. Question: I did not hear any mention of sensitivity testing, which is a common procedure in modeling. Can you identify those parameters which are likely to have the greatest influence on the results of a model? In other words, to what extent have models been subjected to sensitivity testing with respect to their various parameters and processes?

   Answer: To a relatively limited extent. Given the number of processes that are represented in a climate model and the number of variables involved, we have only been able to examine what we believe are the most important cases. Perhaps most sensitivity experiments have been made with respect to ocean-surface temperature, although in terms of the internal parameters of a climate model, clouds may be the more important. Depending upon how clouds are parameterized, the change of surface-air temperature in response to CO_2 doubling, for example, changes by more than a factor of two.

2. Question: Since you are interested in the difference between a standard and an elevated CO_2 level, how big an influence does the initial starting condition have on the difference?

   Answer: If you change the initial condition by a small amount, for example, by changing the last decimal at one point in an otherwise identical calculation, after a few weeks of simulation you will get a different synoptic pattern as a result of the interactions between small and large scales in the model. The time-averaged or climatic conditions, however, will be only slightly changed; such changes are a measure of what is termed natural variability or climatic noise.

3. Question: How well do the models simulate the position of the subtropical highs, and by how much do they change in the future, presumably under increased CO_2?

   Answer: The models do a fair job of simulating the subtropical highs, in that they are all in about the right positions (if we interpret "right" as being about a thousand kilometers). They tend, however, to be too weak, although their seasonal shift is recognizably realistic. The change of the subtropical highs under increased CO_2 appears to be relatively small, as indeed do most changes in the tropics.

REFERENCES

Gates, W. L. 1985. The use of general circulation models in the analysis of the ecosystem impacts of climatic change. *Climatic Change* 7: 267–284.

————. 1987. Problems and prospects in climate modeling. In *Toward under-standing climate change*, ed. U. Radok. Boulder, Colo.: Westview Press.

Kim, J. W., J. T. Chang, N. L. Baker, W. L. Gates, and D. S. Wilkes. 1984. The statistical problem of climate inversion: Determination of the relationship between local and large-scale climate. *Mon. Wea. Rev.* 112: 2069–2077.

Manabe, S., and R. J. Stouffer. 1980. Sensitivity of a global climate model to an increase of CO_2 concentration in the atmosphere. *J. Geophys. Res.* 85: 5529-5554.

Washington, W. M., and G. A. Meehl. 1984. Seasonal cycle experiment on the climate sensitivity due to a doubling of CO_2 with an atmospheric general circulation model coupled to a simple mixed-layer ocean model. *J. Geophys. Res.* 89:9475-9503.

FIVE

Global Climate Change
and California's Water Resources

Henry J. Vaux, Jr.

The prospect of global warming and the consequences of such warming
for California's water resources are both highly uncertain. Still, without
some sense of what these consequences may be, it would be premature
to dismiss the possibility of global warming, however uncertain. California could not have become the most prosperous and populous state
in the nation without adequate quantities of high-quality water. Free-
flowing water also is important in supporting the biological diversity
of the state and in making it an attractive place to live. The continued
availability of water will be crucial in helping to sustain future population and economic growth. A major disruption in water supplies could
threaten both California's economy and the quality of life enjoyed by its
citizens.

California's water resources are highly variable. Two-thirds of the
supply originates in the north while two-thirds of the demand is found
in the south. The state has a distinct two-season climate, with wet winters
and dry summers. Statewide annual runoff averages 72 million acre feet
(MAF) but has varied from a low of 16 MAF to a high of 133 MAF in
recent years. Historically, Californians have successfully offset the loca-
tional, seasonal, and annual imbalance between supply and demand by
developing an elaborate system of impoundment and conveyance facilities, shown in figure 1, which captures surplus flows in wet places and
times and transports them for use in dry places during dry times. Californians have also been able to adapt to the imbalance between supply
and demand by utilizing the extensive groundwater resources that un-
derlie much of the state.

Global warming raises the prospect of significant changes in the timing and magnitude of precipitation and runoff in California, as well as

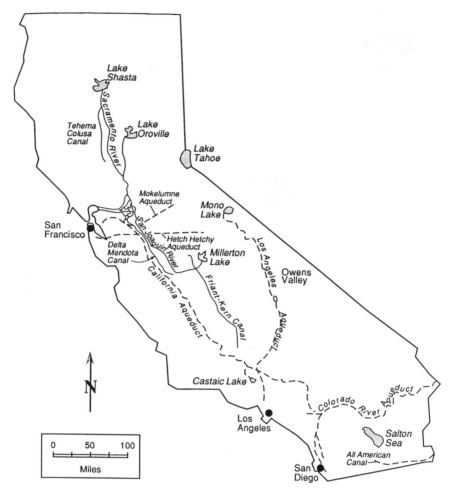

Figure 1. California's major water-conveyance facilities.

the possibility of changes in the patterns and magnitudes of water demands. Many questions emerge as a result:

What are the hydrologic consequences of global warming?

Is the existing physical supply system flexible enough to adapt to the
changing hydrologic circumstances?

To what extent can existing water institutions and water-using behavior be modified to allow the state to adapt to significant reductions in
water supply? What are the costs of failure or inability to adapt?

Considering the costs of failure to adapt, what policies governing the
development and operation of the physical water-supply system, the
institutional arrangements for water allocation and reallocation, and
water-using behavior are justified?

These questions are not independent of one another. The costs of adaptation cannot be estimated without some knowledge of the likely dimension of hydrologic change, yet justification for changes in physical and institutional arrangements for delivering and allocating water cannot be assessed in the absence of information about the costs of failure to adapt. None of these questions can be answered definitively. For some questions, clear answers will require additional research. For others, provisional answers can be developed that will serve both to guide further research and to suggest the kinds of policies appropriate for adapting to global climate change.

This chapter records the deliberations of a group of California water experts about answers to these and other questions related to the impact of global warming on California's water resources. For the most part, those participating in the deliberations believe that the current state of scientific knowledge about global warming and its impacts on water resources is insufficient to permit hard distinctions to be made between short- and long-term changes. Consequently, the ideas discussed here are based on a number of assumptions about specific climatic manifestations of global warming in California, as described earlier in this volume. Ultimately, however, effective public responses to forestall the potentially costly impacts of global climate change will probably depend upon the credible validation of the prospects of global climate warming.

This chapter contains several sections. First, the likely effects of global warming on California's water resources and water-supply systems are identified and analyzed. Second, possible responses to mitigate these effects are enumerated and discussed. Third, the major policy issues are identified. A final section lists recommendations for action and major needs for information.

LIKELY EFFECTS OF GLOBAL WARMING ON CALIFORNIA'S WATER RESOURCES AND WATER-SUPPLY SYSTEMS

Hydrologic Effects

Global warming could produce two quite different kinds of changes with significant effects on California's water resources and water-supply systems. First, fundamental climatic shifts could change the form, timing, intensity, and distribution of precipitation, which in turn would affect the availability of freshwater resources by altering the timing, magnitude, and quality of runoff, and the extent to which it could be captured for use. Second, a rise in sea levels would undoubtedly have adverse impacts on the Sacramento/San Joaquin Delta, the most vulnerable component in California's water-distribution system, and could affect the quality of groundwater in coastal aquifers.

The effect of global warming on levels of precipitation in California is highly uncertain. Even if the most conservative estimates of global climate change are accepted, the likely changes in California's precipitation can only be projected as falling in the range of ± 20 percent. This range is so broad that it is difficult to make any more than a general assessment of the likely impact of precipitation changes alone. Other things being equal, a 20 percent decline in precipitation would create more droughtlike conditions, while a 20 percent increase would create a substantially moister hydrology and enlarge the threat of flooding. Changes in precipitation could also ameliorate or intensify other climatically induced impacts. More specific effects cannot be identified, however, without better estimates of how the timing, intensity, and variability of precipitation might change.

Virtually all analysts agree that increases in average atmospheric temperatures would cause the snow levels in the Cascades and the Sierra Nevada to rise. Exact estimates differ, but the work of Gleick (1987) suggests that snow level would rise by 500 vertical feet per degree (centigrade) of temperature increase. Three degrees of warming, then, would raise the snowline by 1,500 feet. Such a change would substantially reduce the amount of water stored in the snowpack; meanwhile, higher temperatures would cause the snowpack to melt earlier. Estimates of the resulting snowpack loss vary, but the reduction would probably amount to at least 33 percent statewide in an average year, other things being equal. For the Sacramento Basin, the loss would be at least 40 percent; for the San Joaquin Basin, about 25 percent.

Changes in the volume of snowpack and the timing of snowmelt would tend to accentuate California's two-season climate. A smaller proportion of precipitation would be stored in the snowpack, and the earlier melt of the snowpack would cause a distinct skewing of runoff toward the winter and early spring months and away from the late spring and summer months. While the magnitudes of these shifts cannot be estimated with certainty, most studies conclude that, irrespective of changes in precipitation, runoff would increase absolutely in the winter and early spring months and decline absolutely in the later spring and early summer months.

These changes in the timing of runoff could have dramatic effects on the frequency and intensity of floods during wet periods and on the availability of surface-water supplies during the dry spring and summer months, when demand for water tends to peak. California's extensive system of dams and canals, which were built for both flood control and storage of wet-seasons flows, were engineered to permit optimal management of the patterns of runoff observed over the last century or so. A substantial change in the timing of runoff might make it very difficult, if not impossible, to manage reservoir storage capacity to provide accustomed levels of both flood protection and water supply.

A skewing of runoff toward the winter months would substantially increase the likelihood of more frequent and intense flood flows. Such an eventuality could be quite serious, since approximately 75 percent of California's communities contain land susceptible to floods with a current frequency of one in one hundred years. Moreover, existing flood-control structures are designed and operated to protect against floods of this magnitude. Thus, a significant proportion of currently protected urban land, as well as land not now subject to flooding, would become vulnerable to floods.

It should be noted that on a worldwide basis there is some concern about the adequacy of reservoir spillway capacity in regions subject to reduced snowpack and/or increases in winter precipitation. This is not a concern for California, however, since spillway capacities on virtually all major impoundment structures in the state greatly exceed the highest flows ever recorded. Consequently, with proper management, there appears to be no possibility that any major reservoir in California would be lost as a consequence of hydrologic changes brought about by a warming climate.

Responding to changes in the timing of runoff would not be simple. If reservoir operating criteria were changed to make more storage capacity available to manage flood flows, less water could be captured and held for water supply, hydroelectric power generation, and streamflow during the dry season. Extremely difficult choices would have to be made about which water uses to serve and to what extent. These decisions would be further complicated by several other factors. Uncertainties about the extent to which global climate change might affect hydrologic variability could prove particularly vexing. In the absence of definitive information about levels of precipitation, temperature, the size of the snowpack, the time of snowmelt, and the extent to which all of these parameters vary, changes in operating criteria might not have the desired effects.

In addition, the impact of global warming on regions outside of California could play a significant role in defining the terms of trade-off between flood control, water supply, and the maintenance of flows. For example, it will be important to understand the implications of global warming for the Columbia River Basin, since California depends upon power generated there to meet a significant portion of its late spring and summer demand for electricity. Adverse changes in the patterns of runoff in the Columbia Basin could result in major power shortages for California even in the absence of adverse changes here.

Rising sea levels represent a second important potential manifestation of global climate change. These would likely have their own distinct set of impacts. Seawater intrusion already threatens water quality in several important coastal aquifers. Rising sea levels would increase that danger and pose new threats to aquifers in the heavily settled areas of the coastal strip. Rising sea levels would have particularly serious implications for the Sacramento/San Joaquin Delta. Together with adverse wind and tidal conditions, higher sea levels would threaten the already inadequate system of levees that protects Delta islands, and some agricultural lands undoubtedly would be inundated.

In the absence of countermeasures, the loss of some levees, combined with rising sea levels and a decline in runoff during the dry season, would increase saltwater intrusion significantly, threatening the quality of water supplies exported from the Delta. Indeed, particularly adverse combinations of these factors could lead to a complete collapse of the Delta and the potential loss of approximately 65 percent of California's total water supply—and 45 percent of its drinking water supply.

Likely Impacts on Surface-Water Resources and Users

The potential hydrologic changes enumerated above are likely to have significant impacts on California's surface waters. The extent of many of these impacts will depend on the reservoir operating criteria ultimately employed to respond to the increased flood threat. It is unlikely, however, that changes in operating criteria alone could eliminate these impacts altogether.

A skewing of precipitation toward the winter and early spring months clearly would reduce water supplies available for diversion and consumptive uses. Urban water supplies, particularly those of the major population centers that depend on surface water flows, could be substantially reduced. Outdoor uses of water in urban areas would probably have to be curtailed in order to preserve supplies for basic drinking, cooking, and sanitation uses. There would likely be a substantial loss of investment in relatively water-intensive outdoor landscaping. Certain businesses and industries that use water intensively could also be hard hit. Car washes, the nursery industry, and water-based recreational theme parks would be particularly at risk. Outside of agriculture, there should be no widespread damages to industry, however, since California industry already uses water efficiently.

Irrigated agriculture, the state's largest water-using sector, would be subject to substantial water supply shortfalls. Supplies delivered via the Central Valley Project and the State Water Project to irrigated regions south of the Delta would be especially vulnerable. Agricultural surface supplies would probably be reduced more than proportionately, since urban water purveyors would tend to regard agriculture, with its large entitlements, as a supplier of last resort.

Agricultural regions that are solely dependent on surface-water supplies would be particularly hard hit because of their inability to substitute groundwater for surface water. These include some areas on the western side of the Sacramento Valley and the southern San Joaquin Valley, the Salinas Valley, and some irrigated lands in the Sierra foothills. In these areas, irrigated acreage would probably have to be reduced sharply. Such reductions could lead to a domino effect that would threaten the financial viability of irrigation districts that contain marginally productive lands. For districts with substantial outstanding debt on distribution facilities, default on annual payments by a significant number of landholders could lead to default by the entire district. Ironically, this could make additional supplies available to districts with supplemental sources of groundwater or those whose distribution facilities now carry relatively modest levels of debt.

The implications of global warming for in-stream uses are also likely to be serious. Perhaps the most predictable effects would be those on fisheries. Lower flows during late spring and summer are likely to be accompanied by warmer water temperatures. Chinook salmon, in particular, require stable, moderate flows of cold water and are already at the outside limits of their range in California streams. Either flow reductions or temperature increases could be sufficient to extirpate both the spring and winter runs of Chinook salmon. Ocean conditions could also play an important role in determining whether anadromous fisheries survive.

Striped bass would also be at considerable risk, since spring flows appear to be especially important to their spawning cycle. Striped bass could also be adversely affected by reductions in food supply associated with lower flows and increased rates of metabolism caused by higher water temperatures. American shad and steelhead would also be likely to suffer adverse impacts. Conditions in the Delta during April, May, and June are critical for these latter three species, and any changes in Delta flows or salinity could affect their productivity and ultimate survival. Warm-water fishes, by contrast, would probably benefit from lower flows and higher water temperatures. It is possible, therefore, that the mixture of fish species found in California's major streams could be profoundly changed.

The impacts of an altered flow regime on water-based recreation could also be substantial. An increase in the variability of streamflows would reduce fishing and rafting activities by diminishing the time periods during which flows will support these activities. The season for reservoir-based recreation could also be substantially shortened, depending upon how the reservoir pool is managed. In addition, the aesthetic attributes of streams and reservoirs could be damaged significantly during periods of low flow. During periods of high flow, recreation on streams could become considerably more hazardous, particularly on streams where flows are not regulated. Even on regulated streams, the increased difficulty of managing flows with existing reservoir capacity could make downstream recreation more hazardous. Increases in precipitation during the late spring months could help to attenuate these effects; decreases in precipitation would intensify them.

Allocating available streamflow to hydroelectric power generation will pose a difficult problem. During periods of increased streamflow, reservoir storage for power production will have to compete directly with demands for augmented flood control; and during periods of lower flows, with storage to augment fish and wildlife habitat. The result will almost assuredly be some reduction in the reliability of the hydroelectric power generating system. This, in turn, will reduce the capacity of that system to meet peak power demands.

Likely Impacts on the Sacramento/San Joaquin Delta

The Sacramento/San Joaquin Delta, shown in figure 2, is at the heart of California's water-supply system and is the least flexible element in that system. The Delta is not only a critically important source of water supply but is also an important recreational resource, provides habitat for fish and wildlife, and contains 600,000 acres of prime agricultural land. The Delta would be particularly sensitive to climatic warming because it is subject to both the effects of increased streamflow variability and heightened sea levels. Water to support Delta agriculture comes from some seven hundred miles of natural and improved channels. Most of the lands served by these channels are protected by levees, a large number of which are constructed on unstable foundation material and are subject to failure under certain combinations of runoff, tide, and wind. The frequency of flooding is already apparently increasing despite efforts to improve the levees; increases in sea level will increase the incidence of levee failure. The costs of strengthening Delta levees to avert such failure is estimated as high as $3.4 billion (California Department of Water Resources 1982).

A significant portion of Delta agricultural land lies below the prevailing water level. Such land must be drained artificially with sumps and pumps. Heightened sea levels might raise the local water tables, increasing seepage into the agricultural lands and the volume of water that must be pumped out. Drainage pumping lifts also would be increased; pumping lifts for water supply would be correspondingly reduced, however.

The efficiency of irrigation on Delta lands would undoubtedly increase, since low flows would be coincident with the irrigation season. This would lead to reductions in the quantities of tailwater (residual irrigation water) pumped from Delta islands. However, the concentration of pollutants in the tailwater would be higher and, in combination with the reduced flows available for dilution, would lead to further declines in Delta water quality.

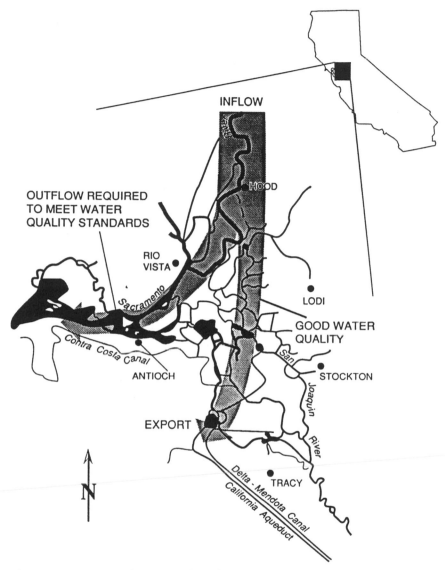

INFLOW

OUTFLOW REQUIRED
TO MEET WATER
QUALITY STANDARDS

HOOD

RIO
VISTA

Sacramento

LODI

GOOD WATER
QUALITY

Contra Costa Canal

San

ANTIOCH

STOCKTON

Joaquin River

EXPORT

N

Delta - Mendota Canal

TRACY

California Aqueduct

Figure 2. Sacramento/San Joaquin Delta.

Low-flow conditions also would increase the time that water remains
in the Delta. Preliminary research suggests that the concentrations of
substances such as trihalomethane precursors increase as a function
of contact time during periods of low flow. Increased concentration of
these potential pollutants could pose serious problems for domestic
purveyors who draw raw water supplies from the Delta.

Additionally, low flows would be accompanied by higher water temperatures, since constraints on reservoir storage would limit cold water discharges. Warmer water would further modify Delta habitats and increase bioactivity, thereby putting some aquatic organisms, including fish species, at risk.

A rise in sea level would increase the tidal prism within the Delta. Intrusion of salinity would be accelerated and the interface between fresh and salt water (the "null zone") would be moved landward. The ensuing effects on water quality would jeopardize both water supplies and fish and wildlife habitat. These effects would be compounded during periods of low flow, when less fresh water would be available to repel saltwater intrusion.

Higher sea levels also would inundate much existing marshland, including the Suisun Marsh, which provides critical wildlife habitat. Although there would be some expansion of wetlands due to higher sea levels, it is not possible to determine whether this would, in effect, substitute for wetlands that would be inundated. The structure and composition of species in Delta wetlands probably would change as a consequence of such substitution, however.

The increased variability in seasonal flows would also have a number of impacts on the Delta. Higher winter flows would increase the probability of floods, particularly when combined with adverse wind and tidal conditions. Increased winter flows would increase the scouring of Delta channels and accelerate the movement of sediments, an effect that could be beneficial. Increases in winter flushing flows would have positive effects on Delta water quality.

The consequences of change in both the quantities and quality of water in the Delta would create the potential for a major disruption in California water supplies. The combination of increases in sea levels and reduced water flows through the Delta during the late spring and early summer months, in particular, could be catastrophic. The increased probability of levee failures, leading to more permanently inundated areas, would accentuate the effects. One estimate suggests that a sea-level rise alone could require as much as 700,000 acre feet of additional supplemental water to maintain Delta water quality (California Energy Commission 1989). Such a requirement would be imposed on the water storage system at a time when seasonal demands for both consumptive and instream uses were peaking. This would enormously complicate the problem of managing reservoir storage for flood control and water supply purposes.

Should the Delta collapse completely, as much as 65 percent of California's water supply could be lost. The portions of central and southern California which rely heavily on imports from the Delta would be particularly vulnerable to such losses. In any case, if a large proportion of limited dry-season freshwater flows must be employed to prevent unacceptable declines in water quality, water may not be available for export. During a period of global warming, managing the Delta could be California's greatest single challenge.

Likely Impacts on Groundwater Resources

California's groundwater resources will be crucial during a period of global climate change because they are the most resilient component of the water-supply system. In addition, the state's groundwater resources are vast, with an estimated one billion acre feet in 394 groundwater basins.

The implications of global warming for groundwater over time are not wholly clear. Areas that have access to groundwater of adequate quality will be able to offset shortfalls in surface-water availability. However, unless these extractions are offset by increases in natural or artificial rates of recharge, the water table will drop, thereby progressively increasing the costs of pumping. Intensified groundwater mining appears quite likely, since reduced applications of surface water, particularly in agricultural regions, will result in diminished rates of recharge. In addition, energy shortages caused by global climate warming could limit energy available for groundwater pumping or make it considerably more expensive.

Assuming, however, that groundwater extractions would have to be increased to mitigate the most damaging consequences of surface-supply shortfalls, the long-term implications for both the quality and quantity of California's groundwater resources still are far from certain. The rates of natural recharge could increase during wetter periods, but it is unclear how local and regional flow fields might change in response to seasonal changes in precipitation and runoff. Areas of recharge and discharge may change so that natural rates of recharge in some localities would increase while others would decline. In some areas the mining of shallow groundwaters could cause upwelling of brackish water with resulting deterioration in the quality of groundwater. In times of surface water deprivation it could become economical to extract water from deep aquifers that would, initially at least, be unaffected by climatic warming. Cost and availability of electricity will be critical determinants of the economic feasibility of such deep pumping. It should be recognized that mining both deep and shallow aquifers extensively over substantial periods of time could lead to subsidence of overlying lands. Significant land subsidence in urban areas could be extremely costly; in agricultural regions it could increase the costs of farming somewhat.

Likely Impacts on Water Quality

Changes in both hydrology and use patterns during periods of global climate warming will adversely affect the quality of California's waters. When runoff and streamflow are heightened, the rates of erosion and sediment production are likely to increase, damaging the biological productivity of watershed lands and aquatic ecosystems, increasing the rate of reservoir sedimentation, and shortening the physical life of many storage facilities.

During low-flow periods the capacity of streams to dilute the concentrations of all types of pollutants will be diminished. In fact, this loss of dilution capacity in surface water is likely to be the most prominent water-quality impact of global warming. Existing water pollution problems will be magnified, and new problems will compound the degradation of water quality.

Increases in water temperatures are likely to contribute to more severe water-quality problems. Warmer water temperatures will result in greater rates of salt pickup by water and increase the concentration of salt in wastewater. Warmer temperatures will also reduce the concentrations of dissolved oxygen and thus decrease the capacity of waterways to assimilate waste. Declines in dissolved oxygen may have implications for aquatic organisms, since the decline in available oxygen will coincide with increases in bioactivity and the demand for oxygen. Warmer water temperatures would also create the possibility of accelerated reaction rates in the water environment, reducing the dissolved oxygen even further. Although the implications of faster chemical reactions in the water environment have not been investigated completely, the presumption is that they would tend to magnify the decline in water quality.

Increases in the rates of evapotranspiration of irrigated crops will make it more difficult to manage salt balances in the root zone. To the extent that additional water must be applied to cropland to maintain productivity, more salt will be introduced into the root zone. In turn, removal of these salts will require additional quantities of water for leaching. These additional leaching requirements would be imposed on a water-supply system that may already be unable to meet the various demands for consumptive and in-stream uses. In addition, the use of minimum leaching fractions will tend to concentrate salts in the drain water. Given that the diluting capacity of receiving waters will be reduced, the problem of managing agricultural drain waters will become even more serious than it already is. Increases in salt pickup and the intensified problems of managing salt balances are also likely to accelerate the degradation of groundwater quality over time.

In summary, there is little question but that water quality will deteriorate as a consequence of global warming. However, a detailed assessment of the likely changes will require more specific information on the hydrologic and human impacts of global warming. Water quality is also likely to vary spatially, so that detailed knowledge of changes in regional hydrologies will be an important prerequisite to understanding specific impacts on water quality and devising means to manage it.

Likely Impacts on Water Consumption

Global climate warming would probably produce changes in water consumption in naturally vegetated areas, in cropped areas, and by the human population. Although a warming of atmospheric temperatures would appear to lead a priori to increased rates of water consumption, what would actually happen is considerably more complicated.

For naturally vegetated areas it is not possible to predict the impact of a warmer climate on rates of evapotranspiration without more knowledge than is now available. The rate of evapotranspiration in plants is influenced not only by wind, humidity, and temperature but also by the amount of soil moisture available to the plant. Thus, in a warmer climate the evapotranspiration of native vegetation could be considerably higher during very wet periods. Also, even with higher evapotranspiration, the rates of deep percolation and groundwater recharge could increase. During dry periods, the rate of evapotranspiration could decline substantially because of a lack of soil moisture.

Existing data show that there is considerable variability in the rates of evapotranspiration under current climatic regimes. Thus, in California, seasonal evapotranspiration for a full grass crop can vary by as much as ± 12 percent. The observed variability during the cooler months of November through February is as high as ± 18 percent. This variability shows that some knowledge of how wind, temperature, radiation, humidity, and soil moisture would change in times of global warming is prerequisite to understanding the likely impacts on evapotranspiration rates in naturally vegetated areas. Although point estimates of evapotranspiration are quite good, much less is known about the behavior of evapotranspiration in larger areas. For example, it is known that the larger the area, the greater the variability in evapotranspiration. Thus, extensive variability in evapotranspiration has been observed across watersheds; but virtually nothing is known about the dynamics that drive these variations.

The projected increase in concentrations of atmospheric CO_2 further confounds efforts to estimate changes in evapotranspiration. Increased atmospheric CO_2 would tend to produce stomatal closure, which, other things being equal, would cause reductions in evapotranspiration. Without more exact knowledge of the changes in temperature, wind, and humidity, however, it is very difficult to predict the extent to which reductions in evapotranspiration resulting from stomatal closure might offset increases in the rate of evapotranspiration caused by warmer temperatures.

Potential changes in evapotranspiration on agricultural lands are similarly difficult to predict because the same variables are at work. In addition, as temperatures and other climatic variables change, growers may respond by shifting to different crop types. Thus, in order to assess probable changes in evapotranspiration on irrigated lands, it would be necessary to know the acreage of different crops and planting and harvest dates as well as the likely changes in climatic variables and soil moisture.

Water demands in the domestic, commercial, and industrial sectors can be expected to increase as temperature increases. Several factors will probably drive these increases. First, California's position on the Pacific Rim suggests that its population and economy will continue to grow. Population growth will put tremendous strain on the state's water resources even without climate change. Although the rate of growth may be reduced from levels that would prevail without global climate change, a net increase in the population of California over the coming decades is likely, and could be very large. Growth in the domestic demand for water will be driven by population growth as well as by increases in water use for landscape irrigation and evaporative cooling. Increased commercial demand for water will similarly be driven.

Likely Economic Impacts

The current information base is insufficient to permit firm conclusions to be drawn about the likely economic impacts of changes in the availability and quality of water due to global climate change. Relatively small changes in climate probably would accelerate existing trends. Thus, the general scarcity of water to meet increasing demands would be intensified. The rates of water-quality deterioration would quicken, further exacerbating the general scarcity. Solving today's water problems would do much to forestall any adverse effects from modest climatic changes.

For more severe symptoms of climatic change, the economic damages are difficult to predict without more precise estimates of probable reductions in streamflow, intensification in the frequency and severity of floods, and the impacts of rising sea levels. Estimates reported in one study suggest that a 30 percent reduction in streamflows could cause direct economic damages amounting to over $200 million annually (in constant 1980 dollars) by the year 2020 (Vaux 1985). This estimate includes only the direct value of lost productivity of water and losses in returns to water purveyors for consumptive uses. It does not include losses from floods, losses associated with changes in in-stream flows, or indirect economic losses. In the absence of any adaptations, these latter losses could be substantial, and it is not unreasonable to speculate that water-related losses in California due to global climate warming could amount to as much as a billion dollars annually.

Continuation of the political stalemate that currently characterizes efforts to resolve California's water problems would constrain the state's ability to adapt to global warming and increase potential economic losses. The stakes in today's water-allocation controversies are perceived to be very high. Yet, if flows are reduced, potential flood damages become much higher, and energy becomes scarcer, the stakes will escalate enormously.

POSSIBLE INTERVENTIONS AND RESPONSES TO IMPACTS OF WARMING ON CALIFORNIA'S WATER RESOURCES

The patterns of successful adaptation to global warming will depend critically on both the rate and extent of climate change. If change is moderate or if it occurs relatively slowly, a series of systematic and deliberate adaptations over time may permit California to adjust at relatively modest cost and without sharp disruptions in activities that depend upon water. If, in contrast, climatic change is rapid and/or large, lead times will be diminished and more substantial adaptations will be called for. There is no scientific basis for predicting whether adaptation to global climate change would be relatively easy and painless or difficult and expensive. Since global climate change and its implications for California are so uncertain, there are compelling reasons to undertake two types of efforts—those which promise to reduce the uncertainty and those inexpensive efforts which will increase the flexibility, resilience, and robustness of California's water-supply systems.

Physical Responses

Of the potential resource impacts, one of the most serious appears to be the seasonal redistribution of runoff toward the winter months and away from the spring months. As noted earlier, this will pose a conflict over whether reservoir storage space should be managed 1) to increase the control of flood flows (which would, presumably, decrease available water supplies for the dry season), or 2) to maximize available water supply and incur the resulting increase in flood damages, which could be substantial. The extent of this conflict has not been precisely characterized although it would be relatively inexpensive to do so. (Simulations of alternative flow regimes can be used to devise optimal operating criteria for reservoir storage.)

If the problem proves to be as serious as it appears, it could be ameliorated to some extent by constructing additional upstream storage. This would increase both water-supply storage and the capacity to control floods.

Managing the Delta in the face of major changes in the seasonality and intensity of water flows, as well as heightened sea levels, is clearly another very serious problem. Here, increased upstream storage would help, but actions would also be required within the Delta. The capacity of some existing bypass systems could be enlarged by raising levees.

Some islands may have to be sacrificed, either to avoid the high costs of maintaining them or to convert them into storage facilities—thereby increasing the storage of surplus flows to repel salt water during dry periods. Widening and deepening some Delta channels would reduce flow restriction in the Delta and make saltwater intrusion easier to manage. Construction of barriers to seal off some regions of the Delta could also help, although barriers would have an adverse effect on Delta fisheries. Some croplands could be phased out. Water flows could be controlled by constructing barriers and storage facilities off-channel.

If efforts to preserve the Delta were to prove too expensive, it would be possible to abandon the Delta altogether. In this event, it would be imperative to develop a conveyance facility around the Delta in order to maintain carriage water to the Delta-Mendota Canal and the State Water Project. If the Delta were not abandoned, such a facility might also aid in water management by reducing the flows necessary to maintain Delta water quality. In addition, it might be designed and operated to maintain at least some of the Delta's critical fish and wildlife habitat.

Heightened water temperatures, with its adverse effect upon fish, could be reduced by retrofitting some reservoirs upstream with equipment to adjust outflow temperatures. This action would affect water temperatures only for a specific distance downstream from the reservoir itself and would probably not have any impact on the water temperature in the Delta itself. In some instances, however, it might reduce the aggregate stress on fish populations enough to make a difference.

Construction of additional storage facilities south of the Delta would probably have little impact on flood control. However, more of the high winter flows could then be stored for use during the drier months, and a larger portion of dry season flows could go to maintain Delta water quality.

Development of less expensive wastewater treatment systems is likely to be especially important during a period of global warming. Such systems could make the recycling of wastewater far more feasible economically and would also allow degraded surface water and groundwater to be treated and used. The development of new cost-effective means of treating water of poor quality will be particularly important in light of the probable reduction in dry season flows and thus in the diluting capacity of streams.

The management of groundwater storage and development of conjunctive-use schemes may be the most important adaptation that Californians can make to changed water regimes caused by global climate warming. Groundwater basins can provide substantial additional storage capacity for California's water-supply system at relatively modest cost.

Underground storage would also minimize evaporative losses, a factor of increasing importance as average temperatures rise.

Conjunctive use and augmentation of groundwater supplies would require construction of more spreading grounds and artificial injection wells, permitting substantial quantities of water from winter flows to be captured and stored. Two factors, however, could constrain the potential of increased groundwater storage and conjunctive use as a means of augmenting water supply.

First, with augmented groundwater storage, the water-supply system could become more energy-intensive. Thus, the potential loss of hydroelectric generating capacity in California and elsewhere could increase the cost of groundwater pumping or severely constrain it.

Second, the fact that groundwater extractions are essentially unregulated in most California basins would almost certainly create major institutional barriers to the increased use of groundwater resources. In the absence of some clear system of groundwater pumping rights or entitlements, groundwater resources will continue to be inefficiently exploited. The fact that the rule of capture currently governs groundwater extractions in most basins means that augmented groundwater supplies cannot always be directed to specific uses and users. It will be extraordinarily costly and time consuming to adjudicate all groundwater basins in the state. Yet, if the promise of groundwater banking for dealing with today's water problems (as well as potential problems associated with global climate change) is to be realized, arrangements will have to be devised to regulate and allocate the state's groundwaters.

Large investments in physical facilities to combat the impacts of global warming are risky. If climate change does not materialize or if its manifestations are relatively mild, these facilities might not be needed—but they would still have to be paid for and would absorb scarce resources that might be more profitably used elsewhere. In the absence of more definitive evidence about possible climate change, proposals for large investments to forestall the effects of global warming should be subject to rigorous probability analyses.

Institutional Responses

The threat of political paralysis over the management and allocation of water resources in an era of global warming underscores the importance of developing water institutions that are flexible and capable of responding to change. Most current institutions, including California's system of water rights, were established in another era, when security of tenure and the need to induce settlement of unoccupied lands were predominant. Time and circumstances have changed; these inflexible institutions are poorly suited for managing water resources today.

As population and the economy continue to grow, existing institutions will become increasingly ill-suited for efficient and equitable management of water. They will be especially inflexible and ineffective if global climate change is superimposed on projected population and economic growth. There are, however, a number of institutional changes that could enhance California's ability to cope with global climate warming. Virtually all of these entail the creation of more flexibility in water allocation and development rules.

The need to develop a rational system of groundwater extraction rights or entitlements has already been mentioned. There is a concomitant need to clarify and improve California's system of surface-water rights. Existing riparian rights and appropriative rights established prior to 1914 have not been recorded. Appropriative rights established between 1914 and 1969 are not recorded as to quantity. There are thus substantial uncertainties in surface-water rights, and these uncertainties will likely become paramount if water becomes significantly scarcer. As with groundwater extraction rights, the process of adjudicating surface-water rights would be extremely costly and time-consuming. If water supplies are substantially reduced, however, individual right holders are likely to seek help in the courts. The prospect of large-scale water-right adjudications during times of substantial hydrologic change could complicate enormously the state's process of adaptation. Moreover, such a prospect could trigger a surface-water counterpart of the "race to the pumphouse" which would artificially stimulate water demands during a time of supply shortfalls.

If adjudication is to occur, it will be far easier and less costly to accomplish when the hydrology is relatively well behaved. In the absence of formal adjudication, procedures will need to be devised and institutionalized for dealing with conflicting claims over water rights relatively quickly and inexpensively. The fact that no action has ever been taken on the Report of the Governor's Commission on Water Rights speaks to the difficulties of rationalizing California's system. There is a strong case for pursuing the Commission's recommendations even in the absence of prospective hydrologic change. The case will be even more compelling if global climate warming occurs.

The resolution of potentially conflicting or overlapping governmental responsibilities could also be important in adapting to global climate change. This is particularly true in the Delta, where differing state and federal responsibilities in the past have constrained the quantities of water ultimately available for delivery to users. The Coordinated Operating Agreement that enhances state and federal coordination in operating the Central Valley Project and the State Water Project may make as much as a million acre feet of additional water available to users. Additional usable water supplies can be developed with further coordination.

Some have suggested that the state should take over the operation and administration of the Central Valley Project in order to increase such coordination. In any case, efforts to coordinate the operation of federal and state projects would maximize the yield of water from the Delta, subject to reasonable constraints on Delta water quality. These efforts should be pursued.

During periods of intensified water scarcity, pressures to reallocate water supplies among various uses are likely to escalate significantly. Current water institutions create substantial barriers to the reallocation of water supplies; they should be modified to permit freer response to changes in the availability of supplies. This can be done by creating water markets that will allow the voluntary transfer of water among different users; or by developing flexible allocation rules that can be quickly and easily applied in an environment of hydrologic change. Obviously, either markets or nonmarket systems will need to provide appropriate protections for third parties who might be harmed by reallocation.

The introduction of voluntary water markets in California would help greatly in adapting to global climate change. Their widespread use would ensure that activities displaced by hydrologic shortfalls are the least valued, and in this way help to minimize the costs of adapting to global climate change. The advent of water marketing would also create incentives for water users to manage scarce water supplies more intensively.

Existing institutions, however, do not provide an obvious means to protect or enhance in-stream uses in the case of widespread water marketing. If the full promise of water marketing is to be realized, institutions will have to be modified to allow for the purchase of appropriated water for in-stream uses. Otherwise, in-stream uses will have no competitive standing in markets; water will be underallocated to in-stream uses; and it is likely that strong political resistance to water marketing would emerge.

Flexible allocation rules could be devised for use in lieu of water markets or to complement such markets. Rules that are both effective and flexible could be difficult to establish, because they would have to originate in an intensely political environment characterized by strong pressures to protect the status quo. Such rules would undoubtedly introduce some economic inefficiencies into the water allocation system but could lead to a distribution of limited water supplies that is perceived to be more equitable than the distribution that would emerge in a market-dominated system.

Changes in water-pricing policies also could aid in adapting to global climate change by providing a relatively decentralized method of regulating demand. Current water-pricing policies require charging water users the average cost of supplying water. These policies are based on historical circumstances, in which the average costs of water supply fall as capacity is expanded. Average-cost pricing policies are also employed to ensure, as required by state law, that water purveyors do not make a profit.

Today, average prices understate the true costs of water supplies and provide false signals to consumers about the relative abundance of water. Moreover, the average costs of water now rise when additional capacity is added to the state's water-supply system. There is thus a case for changing pricing policies to emphasize the marginal costs of supplying water. Marginal-cost pricing policies would cause the price charged to a vast majority of water users to increase. This increase would reflect the higher costs of installing new capacity and provide a more accurate signal to water users about the availability of water.

Marginal-cost pricing policies could be designed to allow water prices to adjust to reflect changes in the relative scarcity of water over time. Thus, during periods of short supply, price might increase quite sharply, causing consumers to further economize and ensuring that water use is brought into balance with available supply.

The strictures on profit making by water purveyors would need to be modified to permit the adoption of marginal-cost water-pricing policies. There are various ways in which this could be accomplished, including a system of rebates or water-pricing schedules that mimic marginal-cost pricing but dissipate the profits by underpricing or subsidizing "lifeline" quantities of water—thereby protecting the poor.

In the absence of changes in water pricing, water could be rationed according to formulas devised to equate use with limited supplies. Under such rules, water-using behavior would be far more restricted than it is now, and activities such as landscape irrigation and car washing would probably be severely constrained or prohibited altogether. Such administrative allocation would have to be centrally controlled, and there would be limits on the flexibility with which the rules could be applied to different classes of consumers. Additional regulatory machinery would have to be established, and water-consuming activities would have to be policed.

Behavioral Responses

By any standard, people who reside in semiarid California are lavish consumers of water. There is considerable room for reduction in levels of water use. For example, domestic per capita rates of water use in the state range from approximately 100 gallons per day to over 300 gallons. By contrast, daily water consumption levels in Israel, which has a similar climate, are now only about 60 gallons per capita.

While water consumers will sometimes reduce levels of use voluntarily in response to some serious but temporary threat such as drought, they cannot be expected to reduce use over longer periods in the absence of incentives to do so. Still, water conservation or economizing behavior with respect to water use will obviously play an important role in determining how well California adapts to the impacts of global climate change on its water supply. There are several relatively simple and inexpensive possible actions that hold promise for modifying water-consuming behavior.

One study concludes that many urban consumers are not even aware of how much water they use or what they pay for it (Bruvold 1988). Furthermore, when provided with this information, consumers tend to reduce their levels of water consumption. Similarly, the majority of agricultural water users are unaware of precisely how much water they use.

For domestic and agricultural users alike, the installation of water meters and the monitoring of water use can be expected to reduce consumption somewhat. Further reductions could be expected in the face of changing prices, as discussed above. Many irrigation districts do not have the facilities required to deliver water on demand. Rather, water is delivered at specified intervals for specified lengths of time. In these circumstances, growers have no incentive to economize on the use of irrigation water—and, in fact, tend to be less aware of how much water they are using. Substantial economies in the use of water could be realized by modernizing the delivery facilities in these districts to permit water deliveries to be made on demand. Modernization, however, will involve investment in storage and conveyance facilities at the district level, and its economic feasibility has not been assessed. Its achievement is likely to vary from district to district, as will the amount of water saved by switching to demand deliveries.

The conservation or economizing of water in all sectors will be a very important component in any effort to adapt to increasing water scarcity. Thus, the design of incentives to encourage conservation and economizing will help in developing an arsenal of responses to global climate warming. Little is known about how people respond to various types of incentives, including price. To the extent that global warming may require very precise management of California's limited water resources, it will be important to learn in some detail how and why people respond to different types of incentives.

CENTRAL QUESTIONS ABOUT RESPONSE CAPABILITY

In the absence of more definitive information about the timing and intensity of global warming and its impacts on water resources, it is very difficult to assess which actions will be feasible and which will not. Policy changes and actions that might be unthinkable during more normal times often become attractive during times of crisis. Therefore, a number of key questions will need to be addressed in considering whether an effective response can be mounted to global climate change. These questions relate directly to the feasibility of the various responses described in this section.

It seems clear that no single response, however feasible, will be fully adequate to cope with global climate change. A massive investment in new impoundment and conveyance facilities is unlikely to be feasible, simply because of the cost. In a period of significant climate change many new demands for public investment are likely. In addition, some economic dislocations will probably be unavoidable. The capacity of the state to underwrite large new investments in the state's infrastructure during significant climate change is likely to be quite limited. Moreover, the federal government may also be constrained in the level of support which it can offer, since climate change will likely have adverse impacts throughout the nation.

Physical facilities are not only costly but are also known to cause ecological damage. There is thus at least the possibility that new physical facilities could compound some adverse ecological impacts caused by climatic warming—as well as ameliorate others. Nevertheless, some additional physical facilities undoubtedly will be required. In its current condition, the Delta probably cannot survive a combination of higher sea levels and an intensified two-season pattern of runoff. Thus, some combination of additional upstream storage and in-Delta facilities will be required if the Delta is to be stabilized. Alternatively, it may be more prudent to abandon the Delta, either partially or entirely. In this case, new facilities will be required to convey water around the Delta, or portions of it, in order to preserve the critical capacity to export water to central and southern California. Beyond this, it is not clear what role physical facilities might play in ameliorating climate-induced change in the water system.

One key set of questions, then, is: **To what extent can the construction of new physical storage and conveyance facilities protect Californians against the adverse effects of climate change on their water-supply system? What would the costs of those facilities be? Can state and local government afford to make the investment, given the competing demands in other sectors of the state's economy? What would the ecological impact be? To what extent would adverse ecological impacts associated with new construction be offset by the preservation of habitats and species that would result from the new facilities?**

Key questions about economic feasibility relate to both the efficiency and equity effects of various responses to a changing water situation. Other things being equal, the state and its citizens would not want to pay more to respond to global climate change than absolutely necessary. Yet, other things are not equal; and it matters very much who pays and who benefits. While equity issues will almost certainly be critical in designing a politically feasible response to global warming, equity is often attended to without regard to its costs.

Current preferences, for example, emphasize many of the intangible environmental and amenity values associated with free-flowing streams. If water availability is reduced over the long term, in-stream flows may have to be sacrificed to some extent. It is clear that water uses with tangible economic payoffs no longer dominate water-allocation decisions in California. Yet it is not clear what economic values are being lost as a consequence of efforts to protect the environmental and amenity values of free-flowing streams. Decisions about how to respond to global climate warming will be easier if there is widespread agreement about both the worth and the costs of amenities and environmental values associated with in-stream flows.

Among the key economic questions are: **How is the cost burden of adapting to global climate change to be distributed? How are the benefits of adaptation to be distributed? How do these distributions change if the response is weighted heavily toward the construction of new facilities? How do the distributions change if the response is weighted heavily toward institutional reforms such as water marketing and water pricing? What levels of investment are needed to develop and sustain different levels of firm water supply? How much can state and local government afford to invest in the water systems? What values are associated with in-stream flows? What balance should be struck between in-stream flows and consumptive uses of water in a time of extreme shortfall in runoff?**

The feasibility of making the institutional changes necessary to cope with global climate change depends on the perceived payoffs. The payoffs from such changes are likely to appear much higher when a crisis is imminent than when it is merely an uncertain event. The timing of institutional change, therefore, could be critical. Moreover, such change will be more difficult when certain groups perceive that they will be harmed by it. In any case, global warming will be far easier to respond to if a set of relatively flexible water institutions are in place at the time of its onset; but this may be very difficult to achieve.

The major questions related to the feasibility of institutional change are: **Will appropriate institutional change be feasible in the absence of hard evidence about the water-related impacts of global warming and hard evidence indicating that such warming is likely? Are there ways of recording and improving California's system of surface-water and groundwater rights that will be less costly and time consuming than adjudication of most groundwater and surface-water basins? To what extent can government continue to indemnify those who may be harmed by institutional change?**

MAJOR POLICY ISSUES

A number of major policy issues will have to be addressed in fashioning a response to the possible adverse impacts of global warming on California's water resources and water-supply systems. The prospect of global warming is uncertain. The specific effects of global warming on California's water resources are even more uncertain. If those effects are substantial, it is also uncertain whether California will have the resources either to mitigate them or to indemnify everyone adversely impacted. Risk and uncertainty can be reduced by research aimed at better understanding of 1) the consequences of a doubling of CO_2 in the atmosphere, 2) the effects of that doubling on precipitation and runoff in California watersheds, 3) the effects on water consumption, and 4) our ability to adapt existing water-supply systems to accommodate those effects. However, no research program can eliminate risk entirely.

California's water facilities have never been designed to protect against all droughts and all floods. Different responses and different levels of response to the impacts of global warming on water resources inherently carry with them different levels of risk. Changes in hydrology and in hydrologic variability are unlikely to be linear, so past records will not be particularly helpful in anticipating them. As a consequence, some risk will be associated with any strategy employed to manage the water-related effects of global climate change. The challenge is to decide what the acceptable level of risk should be. Moreover, in the absence of information about costs, less risk will always be preferred to more risk. Thus, realistic and informed decisions can be made only if the costs, both monetary and in-kind, of different levels of risk are clearly delineated. This is one major policy issue.

Within certain limits, institutional change and changes in water-using behavior can be substituted for capital investments in impoundment and conveyance facilities. For example, the inefficiencies inherent in California's existing water-supply system can provide a buffer against declines in water supplies and water quality; but institutional changes such as creation of water markets or imposition of different water-pricing rules will be required to take advantage of the buffer. Similarly, changes in water-using behavior which economize on water use can help stretch limited water supplies and reduce the deterioration in water quality. Such institutional and behavioral responses typically do not require the investment of capital resources. They are not costless, however. Accustomed patterns must be broken and the uncertainties inherent in any change must be dealt with. A second major policy issue, then, involves selecting the mix of institutional, behavioral, and physical responses to the adverse effects of global climate warming.

If global climatic change does occur, government is not likely to be able to indemnify all those who will bear its costs. Policies that are developed to respond will determine not only the extent of indemnification but also who gets indemnified. This is a third major policy issue. Most historical experience with public policies has focused on how to distribute governmental largess. Significant changes in California's water supply will require government to focus on the very different question of how to distribute the costs of change.

The issue of how costs are to be distributed involves a host of more specific questions. What are the terms under which environmental and amenity uses of water should compete with consumptive uses that have readily measurable monetary payoffs? What weight should be given to environmental protection? What weight should be given to supporting additional economic and population growth? To what extent should communities or individuals who suffer directly from specific water-reallocation decisions be indemnified, and how should they be indemnified? To what extent should third parties who are damaged by response strategies be indemnified? To what extent should considerations of equity or fairness be permitted to override considerations of efficiency?

The effectiveness with which these policy issues are resolved will depend partly upon the information available about the likely manifestations of global warming on California's water resources and the nature of the various options available for responding. The costs of adapting to the effects of global climate change on water resources can be significantly reduced if those effects can be anticipated with a minimum of uncertainty and with adequate lead times to permit reasoned responses to be developed.

RECOMMENDATIONS AND MAJOR NEEDS FOR INFORMATION

Hydrologic Impacts

Additional data are needed on the nature of likely impacts of global warming on California's water resources.

- Studies are needed to develop improved understanding of the interrelated physical and biological processes affecting the hydrology of watersheds under altered climatic states. The ways in which these processes vary among watersheds in different regions of the state also need to be understood.
- Research is needed to enhance the understanding of a) the complex relationship between aquifers and atmospheric conditions, b) the relationship between runoff and deep percolation, including the role of artificial recharge, and c) the determinants of groundwater quality.
- Research is needed to assess the response of both wildland vegetation and irrigated crops to global warming. The issue of whether increased stomatal resistance caused by higher-atmospheric carbon dioxide would offset the effects of higher temperatures on evapotranspiration is particularly pertinent.

Water Quality

The impacts of global warming on water quality are likely to be substantial. Devising strategies to overcome further declines in water quality is complicated because current water-quality monitoring is inadequate. Existing information on water quality is insufficient to permit existing trends to be identified.

- Priority should be given to a) additional routine monitoring of water quality throughout the state, and b) research to develop improved water-treatment processes.

Management Strategies for the Delta

The Delta is the most crucial and perhaps the weakest link in California's vast water-supply system. Information is needed on whether the Delta can be stabilized and managed during times of altered patterns of runoff and seawater intrusion, and on the costs of doing so. Additionally, information is needed about how the water-supply system could be managed in the event that the Delta collapses or cannot be stabilized.

- Improved means of conveying water around and/or through the Delta may be necessary to reduce the vulnerability of this pivotal component of California's water-supply system.
- Changes in operating rules for reservoir releases and export pumping should be investigated to ascertain the extent to which the existing system can be adapted to global climate change.
- The possibilities for storing water within the Delta should be evaluated.
- The costs and benefits of additional upstream storage for carriage water should be assessed.
- The biological and ecological impacts of alternative regimes for managing the Delta and its associated watersheds should be thoroughly investigated.

Management of Water Resources

The possibility of adapting to global warming by modifying both water institutions and water-using behavior needs to be better understood. In addition, the potential economic consequences of global warming and various alternatives for adapting to it need to be identified and reported.

- Effective institutions for the management of groundwater resources need to be developed.
- Research is needed in the following areas: a) studies of the comparative value of reservoir capacity for storage of water supply and flood control, b) studies of the comparative values of water in different uses, including in-stream uses, in times of acute water scarcity, and c) studies of how people respond to different types of water conservation programs.

• Systems of incentives that promote efficient water use in all sectors need to be developed and implemented.

A FINAL RECOMMENDATION

Today California's water economy is not in good shape. Even in average years, water supplies are balanced with water use only because of groundwater mining, a practice that cannot be sustained indefinitely. Supply and demand are brought into balance in successive drought years only through mandatory water rationing and similar ad hoc policies. The California Department of Water Resources projects that the disparity between supplies and demand (at current prices) will continue to increase as California's population grows unless new facilities and new policies are developed. Perhaps more significantly, the gap between supply and use is typically reckoned only in terms of consumptive uses. When the environmental and amenity uses of in-stream flows are considered, the gap becomes even larger.

The quality of California's surface waters and groundwaters is deteriorating inexorably. Toxic wastes, residues from irrigated agriculture, and shortsighted watershed management practices all threaten to reduce water quality even further. Efforts to control existing threats of water contamination have not been fully successful, and effective strategies for dealing with future problems have not been developed. The continuing degradation of California's waters threatens to widen even farther the disparity between available supplies of adequate quality and projected water demands. By permitting the degradation of water quality to continue, Californians contribute to a worsening of the future water-supply situation as surely as if they destroyed existing water-supply facilities.

The fact that the response to the current and future water situation can be characterized largely as political paralysis is disturbing. The capacity of the state to solve its water problems appears to be diminishing even as those problems grow. Virtually all of the means proposed for dealing with today's water problems are strongly opposed in some quarter.

The failure to solve current water problems will magnify the difficulties of responding effectively to the additional problems that would accompany prospective global climate change. Since there is much uncertainty about both the dimensions of global warming and the impacts of such warming on California watersheds, an immediate and prudent response would place high priority on solving today's water problems. Such a response would include 1) building more flexibility into the existing water allocation and delivery system; 2) developing incentives that promote economically efficient levels of water use; and 3) developing institutions to better manage the quality and quantity of groundwater resources.

Simultaneously, it will be important to monitor for signs of climate change. Continued efforts must be made to develop baseline information on climatic, hydrologic, and ecological trends, and to improve facilities for maintaining, analyzing, and distributing these data.

REFERENCES

American Association for the Advancement of Science. 1989. Climate and water. In *Report of the AAAS Panel on Climate Variability, Climate Change and the Planning and Management of U.S. Water Resources*, ed. Paul Waggoner. Washington, D.C.: Climate Project, AAAS.

Bruvold, William H. 1988. *Municipal water conservation*. Water Resources Center Contribution #197. Riverside, Calif.: University of California Water Resources Center.

California Energy Commission. 1989. *The impacts of global warming on California: Interim Report*. Sacramento.

California Department of Water Resources. 1982. *Delta levees investigation*. Bulletin 192-82. Sacramento.

Gleick, P. H. 1987. Regional hydrologic consequences of increases in atmospheric CO_2 and other trace gases. *Climatic Change*, vol. 10.

Revelle, Roger. 1988. The future of southern California water supply. In *The water cycle: A symposium*. Scripps Institution of Oceanography Reference No. 88-18, December. San Diego.

Riebsame, William E., and Jeffrey W. Jacobs. 1988. *Climate change and water resources in the Sacramento-San Joaquin region of California*. Natural Hazards Research and Applications Information Center, Institute of Behavior Sciences, Working Paper # 64. Boulder, Colo.: University of Colorado.

Rosenberg, Norman J., William E. Easterling III, Pierre R. Crosson, and Joel Darmstadter, eds. 1989. *Greenhouse warming: Abatement and adaptation*. Washington, D.C.: Resources for the Future.

Vaux, H. J., Jr. 1985. Some economic implications of climatically induced water supply deficits. Paper presented at symposium, Response of Water and Aquatic Biota to Increases in Atmospheric Carbon Dioxide. Annual Meetings of the Ecological Society of America and the American Society of Limnology and Oceanography, at University of Minnesota, Minneapolis.

SIX

Global Climate Change and California Agriculture

Lowell Lewis, William Rains, and Lynne Kennedy

No discussion of potential climate change in California is complete without an assessment of its implications for agriculture. Agriculture is a premier contributor to California's economy, but it is among the leading industries likely to suffer from the effects of global warming. It is imperative, therefore, that researchers and policymakers begin now to assess the stresses that may be placed on today's agricultural system by climatic change.

The importance of agriculture to California's economy is demonstrated by marketing receipts of over $15 billion annually. Estimates of the total contribution of agriculture and its related industries suggest that between 11 and 20 percent of the state's gross product comes from this sector. While less than 3 percent of the population is directly engaged in farming, one in every five wage earners is employed in an agriculture-related field.

California agriculture contributes to the health of the national economy as well as to the welfare of the state's residents. California produce valued at $4 billion is exported each year; over half of this revenue comes from the nations of the Pacific Rim. California is the only state in the nation that produces commercial quantities of almonds, walnuts, pistachios, nectarines, olives, dates, and prunes. California is also the largest producer nationally of more than twenty other agricultural commodities, including such crops as sugar beets, processing tomatoes, strawberries, lettuce, peaches, grapes, and navel oranges, and it ranks second in dairy products and cotton.

Even apart from its economic importance, however, California agriculture offers significant social benefits for many residents. Although the state's current population is more than 90 percent urban—second in the nation only to New Jersey—Californians in the eighties are moving to

97

the "country" in increasing numbers to enjoy a more open and unfettered life-style. Many of these urban emigrés run small specialty or organic farms in addition to their other employment; such mini-farms now account for more than 30 percent of all farms in the state. These new rural residents are joined by emigrés of another sort—those fleeing adverse conditions in their homelands—who have found opportunity in California's fertile valleys.

PERSPECTIVES ON CLIMATE CHANGE

The Impact of Global Warming on Agriculture

Economists have long recognized the differences that distinguish agriculture from other industries; agricultural economics is, in fact, a distinct discipline. Unlike banking or automotive manufacturing, farming is constrained by seasons and weather as well as by market forces. Additionally, the upper bound on the amount of land available for agriculture is fixed, and not all land is equally suitable for all types of production.

Agriculture has traditionally been a risky business, but in California the relatively predictable climate and ability to irrigate have provided a competitive edge to production of many of the state's agricultural commodities. The wine-growing areas of France, for instance, are famous for vintage years, because a good year is exceptional. California's traditionally dependable climate has tended to erase the vintage-year concept for all but the most dedicated connoisseurs of California wines.

Under global warming, however, temperatures will generally increase; amount, distribution, and timing of precipitation may change; and land area suitable for agriculture could increase or decrease depending on the region. The importance of these factors to agriculture means a magnification of the effects of global warming for the agricultural sector. Output by competitors and demand by consumers will change according to shifts in the comparative advantage of various regions for production of specific goods, and some cropping areas may require relocation. Virtually the entire environment in which agriculture is now conducted may change. A period of transition from one climatic equilibrium to another will probably be accompanied by an increase in the variability of temperature, rainfall, and other natural elements.

As is true for the wine industry, the loss of a consistent and predictable climate due to an increase in the variability of weather will bring a corresponding increase in the risk associated with much of California agriculture. Thus, agribusiness will have to pursue new strategies: entrepreneurs and corporations will need to further diversify their agricultural holdings, and managers will have to develop new cropping systems to increase management flexibility. Makers of public policy may need to change government subsidies and crop insurance programs to address new situations.

Present Trends Extrapolated

Important though the changes brought about by global warming may be for tomorrow's productivity, stresses that already jeopardize California agriculture today may prove more important still. If the diverse challenges that currently confront agriculture are not met, they will be exacerbated by the effects of global warming. A consensus of many scientists is that agriculture is on a collision course with itself; without significant change there may be no agriculture left in some parts of California for climate change to affect.

Of the problems facing California agriculture today, the need for adequate quantities of quality water is one of the most pressing. Traditionally, this need has been met by irrigation. Unfortunately, however, irrigation waters in some areas of the state are presently causing an accumulation of salts in quantities detrimental to crop yields and survival. Unless irrigation practices improve and leaching and proper drainage occur, many of these areas will lose their productive capacity in a matter of decades, with or without global warming.

As in most of the developed world, California agriculture relies for high yields and unblemished products on chemical inputs in the form of fertilizers, pesticides, hormones, and antibiotics. In recent years, however, public awareness has grown that these chemicals can have long-term effects on the environment and the consumer. Governmental regulation and consumer anxiety are continually reducing the low-cost options available for combating pests and diseases and assuring adequate plant or animal nutrition.

Increasing air pollution presents yet another threat to California agriculture. Particularly in the Central Valley, the state's burgeoning population has increased demand for housing and transportation. Rising smog levels there already evidence the enormous potential for air-quality problems in the interior valleys. Surrounded by mountains, these valleys serve as natural collection basins both for locally generated pollutants and for those blown in from coastal population centers such as the San Francisco Bay area. Damage equal to a 20 percent reduction in yield has already been documented for several crops in the San Joaquin Valley.

A Rational Strategy

For California to maintain its present level of agricultural production, far-sighted researchers and policymakers must consider which areas of technology and public policy to target for improvement. Resources should be directed now toward the resolution of existing problems and the mitigation or prevention of further problems caused by climate change. To the extent that current efforts are successful, farmers will be able to make strategic management decisions that may reduce the potential societal and environmental costs of these problems.

At present, a key limitation to efforts to adapt agriculture to future conditions is the high level of uncertainty regarding the anticipated changes. Existing climate models are unable accurately to predict parameters of importance to agriculture such as amount and seasonal distribution of rainfall, diurnal and seasonal extremes of temperature, and microclimate responses in particular regions. Thus, agricultural researchers are left with only one tenable option: they must increase the *resiliency* of agricultural systems and be on the alert to recognize symptoms of change as they occur.

Increasing agriculture's resiliency will mean expanding the number of management options available to farmers; reducing dependency on chemical inputs; enhancing biological diversity and genetic tolerance to stress; and improving monitoring and information systems available to farmers, researchers, and policymakers. These strategies can be undertaken through better understanding of the requirements of crop species and animals as well as through intelligent consideration of agriculture's appropriate role in ecological, socioeconomic, and political systems.

INCREASED TEMPERATURE AND CO_2 CONCENTRATIONS

Global warming will entail complex changes in the physical environment upon which agriculture depends. The implications of these changes for crops and livestock will also be complex. In the following sections, the effects of changes in temperature, CO_2, and precipitation are first treated separately, then combined in sections that reflect a more systems-level approach.

Increased Temperature

Negative Effects on Crop Physiology.   For biological systems the most important effect of increased concentrations of greenhouse gases in the atmosphere will be the rise in global temperatures. Most of the physiological processes involved in the development of organisms are driven by temperature: development occurs within a range between low and high lethal temperatures, and the rate of development within this favorable range is directly, if not linearly, related to temperature. Beyond this favorable range, incremental increases in temperature cause rapid decreases in growth rates until, finally, mortality occurs. An increase of two to four degrees Celsius could therefore have important implications for many California crops.

Along with average temperatures, short periods of intense heat occurring during critical times in the growth cycle of a susceptible plant can have negative results. Short periods of high temperatures can retard morphological development, induce male sterility in flowers, reduce grain fill or fruit development, and cause fruit drop. Because the variability of weather patterns is expected to increase along with average temperatures under climate change, damage due to extreme temperatures may be an increasingly severe problem.

The number of crops that could be affected by increased average and peak temperatures is extensive. Virtually every category of crop grown in California includes species that are susceptible to temperature changes in the realm of those predicted under a global warming scenario. Pears, kiwi, apples, oranges, cherries, most leafy vegetables, cotton, and melons are just a few of the many crops at risk.

California's desert valleys already experience some of the highest summer temperatures of any major crop production zones in the world. In the Imperial Valley, cotton, okra, sudan grass, alfalfa, and sorghum consistently produce economic yields, but many cultivars produce suboptimally owing to high temperatures. Of the cropping systems found in the southern San Joaquin Valley, many are not successful in the Imperial Valley, owing in part to a three-degree (Celsius) difference in average temperature. If average temperatures shift upward, agricultural production in the Imperial Valley could be greatly restricted, and parts of the San Joaquin could approximate current conditions in the Imperial Valley.

In addition to average and peak temperatures, other heat-related variables can limit crop growth; low temperatures and diurnal variation (the difference between day and night temperatures) are important to some species. Most deciduous fruit and nut crops in California have a chilling requirement that must be met to ensure an orderly, compact bloom period in spring. Many of these same crops require a certain spread between night and day temperatures. When this is not met, yield and fruit quality may suffer.

Warmer temperatures may affect the incidence and type of pest populations occurring in a given region. The pink bollworm, for example, has ravaged cotton in the Imperial Valley but has never been established in the San Joaquin Valley, although bollworm moths are consistently found there. Researchers believe that the San Joaquin Valley may have escaped serious infestation owing in part to inability of the bollworm to survive the colder, longer winters there. In the San Joaquin Valley, autumn light and temperature conditions do not provide the right signal to the bollworm to begin storing energy needed to enter diapause, the insect equivalent of hibernation. Because of this (together with a sterile moth release program and other management practices), relatively few of the moths survive the winter to emerge in the spring, and populations stay naturally low. Scientists fear that an increase in average temperatures could allow the bollworm to better survive the winter in the San Joaquin Valley.

Increased temperature can also cause a change in the peaking of pest cycles with respect to the life cycles of host crops. Lygus bug, another insect known to devastate cotton crops, overwinters in the hilly rangelands of California and in the mature crowns of alfalfa plants. Those bugs that overwinter in grasslands migrate to irrigated crops such as cotton when the grasses dry in late spring. If this migration coincides with the reproductive phase in the development of the cotton crop, serious yield losses can result. High temperatures, together with late spring rainfalls, could result in earlier maturation of the cotton crop and a later migration of lygus bugs, increasing the likelihood of infestation at the reproductive stage.

Implications.   Potential responses to the types of problems described above will likely be variants on three basic strategies: 1) genetic manipulation through plant breeding and biotechnology, providing better tolerance to environmental stimuli; 2) management strategies changed to optimize production under new constraints; or 3) cropping patterns shifted to follow the movement of climatic regions. The potential for breeding crops tolerant to heat is great, although few resources have been devoted to this effort to date. Likewise, few resources have been devoted toward understanding and improving field management at the cropping systems level, and the potential for finding adaptive improvements is high. The final option, shifting cropping patterns geographically, will require higher farm investments during the transition period and will probably impact the total acreage sown to various crops. A few species could actually be eliminated from California, while others might be successfully grown for the first time. Although it is tempting to think of regions of adaptation for crops as simply moving north along the new isotherms, diverse geology and geography will likely interpose microclimate features that reduce the success of such migrations. There are no Salinas or Napa Valley equivalents, for example, north of Santa Rosa, extending as far north as Washington State. Soil types, which can vary widely within a small area, will provide additional complexity to the crop-adaptation puzzle.

Photoperiod requirement (number of hours plants are exposed to darkness) is another variable that might constrain man's ability to shift cropping regions latitudinally. Although genetic improvement of many crops has brought virtual day-neutrality, the adaptation of other varieties to regions where the cropping season is too long or too short has included use of photoperiod response as a mechanism to control days-to-maturity. For these varieties, as well as new species that might be introduced to California, day-length requirements could affect the short-term potential for geographic shifts in cropping patterns.

Positive Effects on Crops.   Not all of the changes due to global warming will be negative. Although significant dislocations will occur in the short run if climate change is rapid, long-term results could be favorable when (and if) the climate stabilizes at a new equilibrium. Temperature increases in southern parts of the state could allow for introduction of tropical species, and winter vegetables and avocados, which are currently limited to southern California, could become viable in the San Joaquin Valley. More double cropping could be accommodated owing to warmer winters, and some pest species might be inhibited, rather than helped, by higher temperatures. Interactions of microclimates with crop species, while limiting distribution of some types, would also assure that many specialty crops will continue to be grown in the state.

Impact on Animals. Increased temperature can have both direct and indirect effects on livestock. Indirect effects result from changes in crop production zones and nutritional value. For example, alfalfa loses quality in hot weather through increases in cell wall content, meaning that additional feed is required to meet the nutritional and caloric needs of ruminants that consume the alfalfa. Because most animals raised for dairy and meat production are homoiothermic (warm-blooded), they are highly adaptable to environmental changes within the ranges assumed for global warming. Nonetheless, high temperatures can cause molt and mortality in poultry, reduced milk production in dairy cattle, and reproductive inefficiencies in swine. Feedlot cattle do not gain weight as rapidly during prolonged periods of elevated temperatures as during cooler periods.

Fish, which are poikilothermic (cold-blooded), may be the most influenced by temperature change of any food animals. Salmon, steelhead, and other species that need cold water to survive are the most at risk. Decreases in streamflows together with increased ambient temperatures will result in warmer water. If these fish are to survive, greater releases will be required from reservoirs in order to keep stream temperatures at acceptably low temperatures. Warmer water will also contribute to development of algal blooms and increased biological oxygen demand in fishponds (reducing the oxygen available for fish).

Just as plant pests will be affected by global warming, the distribution of some livestock pests may change with warmer temperatures: culicoides, mosquitoes, and other pests may move north, increasing existing problems and creating potential new problems. Some of these pests carry important infectious diseases of sheep, cattle, horses, or humans, such as the bluetongue and encephalitis viruses. New strains of these viruses as well as viruses from subtropical climates may appear in California. Warmer climates will increase the length of the seasons that some of these vector-borne diseases are active, resulting in a need for more extensive control programs and consequent costs to producers and the environment.

Possible responses to protect livestock from the effects of excessive temperature are likely to be costly. Livestock may be bred for greater heat tolerance: those animals with greater surface area per weight and greater capacity to perspire might be favored over larger species. However, such breeding efforts will require time. Many of the smaller breeds of heat-tolerant animals, for instance, are inefficient milk producers. In addition to improvement of current California livestock, new breeds from tropical climates such as Brahman cattle could be adapted and improved for California production. Where breeding efforts are ineffective or time-consuming, livestock that are not adapted to increased temperatures may be raised in altered environments. This may be accomplished by changing the location where the animals are raised, or by building shelters to maintain them.

Increased CO_2 Concentrations
Physiological Effects on Crops.   According to some scientists, present-day crop species evolved under significantly higher concentrations of atmospheric CO_2 than exist today. If true, this helps explain why availability of CO_2 is limiting under most conditions for current crop species. The effects of increased CO_2 concentrations on crop development and production have been studied, giving the following rule of thumb. Assuming other factors nonlimiting, a twofold increase in atmospheric CO_2 will raise the rate of photosynthesis so that a 50 percent increase in plant biomass might result. The partitioning of this increased mass through the plant is fairly uniform, making the estimate valid for harvested portions of the plant. However, variation in response to CO_2 exists between different crop species. Legumes, normally in an energy-deficit state due to the requirements of nitrogen fixation, may reduce their deficiency through the effects of increased CO_2 on photosynthetic rates. Positive effects for tree crops may be small, owing to the many nonphotosynthesizing parts of the tree, such as shaded leaves, trunk, and branches. C_4 plants (those like sorghum and corn which use four carbon acids to fix CO_2) will lose some of the advantage they presently have over C_3 plants (plants that use three carbon acids to fix CO_2). However, this relative loss in efficiency by C_4 plants may be compensated for by increased water-use efficiency; trials on sorghum under increased atmospheric CO_2 have shown such an increased efficiency of plant-water use. Acting to counterbalance this effect is the increase in plant-leaf temperatures that could result from partial stomatal closure—the primary factor contributing to increased water-use efficiency.

Although plant biomass may increase in response to higher concentrations of atmospheric CO_2, that mass may have a modified chemical composition. Carbohydrate content is likely to increase disproportionately with respect to nitrogenous compounds. Pest species could respond to this change by eating more plant. Soybean looper (*Pseudoplusia includens*) larvae fed soybean foliage grown at 650 ppmv CO_2 consumed 80 percent more than did larvae fed on leaves grown at 350 ppmv CO_2. In other experiments, plants grown under higher CO_2 levels had heavier infestations of aphids than did controls. The effects seem to be insect- and plant-dependent, however, making generalizations difficult. Additionally, the short duration of the studies has not permitted buildup of insect predator and parasite species, so it is difficult to predict the net long-term effects of increased and modified biomass on pest infestations.

Impact on Livestock. Just as insects may need to increase their intake of plant matter to compensate for reduced nutritional value, livestock may need to eat more, or at least eat a more diverse diet. No direct metabolic effects of increased CO_2 are expected for animals, since the partial pressure of oxygen is not expected to drop below necessary levels.

Combined Effects of Increased Temperature and CO_2

Unfortunately, little information is available on the combined effects of increased temperature and CO_2. It may be assumed, however, that although response by individual species will differ, increased rates of maintenance respiration could offset to some extent the benefits to plants of elevated CO_2 levels.

EFFECTS OF CHANGING WATER QUALITY AND SUPPLY

Projected Changes in Precipitation

Climate models unfortunately give little insight into either the magnitude or the direction of expected changes in precipitation. Drought or flood could both be outcomes of rainfall that is "plus or minus 20 percent" of present averages. Projections regarding available stored water, however, are somewhat stronger than those regarding total precipitation. Owing to higher temperatures, proportionately less snow and more rain will likely fall in the Sierra Nevada. This means that less water will be stored as snowpack, and more runoff will occur during winter and spring. If the total amount of both precipitation and reservoir capacity remains constant, less water will be available in summer and fall from stored supplies.

Changes in the water potentially available for agriculture could result from other climatic factors. If the models are correct and shifts in the timing of the rainy season occur, the first rains of fall may come later, and spring rainfalls may extend later into the year. Variability in the timing of rainfall could increase also, as part of the general trend toward greater climatic variability.

The Current Water Situation

Changes in precipitation and total stored water may prove ultimately more important for California than any other feature of global warming. Even without climate change, water is expected to be the limiting resource of future decades. Supply is not expected to keep pace with demand, and agriculture will be in an increasingly vulnerable position to lose some of its allotment.

Demand. Total commercial, public, and private consumers presently use a net 34.2 million acre feet (MAF) of water per year in California. Of this water, groundwater supplies 23 percent of demand, with an average annual overdraft of about 2.0 MAF. Surface waters contribute the rest. The supply of water is normally adequate for California's needs, with the exception of those areas such as the San Joaquin Valley, where overdraft of groundwater is a consistent problem.

By 2010 the state's population is expected to increase by 38 percent. Municipal and industrial demand for water will grow by almost 29 percent, and wildlife, recreation, and energy uses will consume 10 percent more than at present. By contrast, agricultural demand for water is predicted to remain relatively stable.

Supply. In contrast with the increasing total demand for water, the supply is likely to remain constant, assuming no changes in current precipitation. Only one site has been approved for construction of additional surface storage, and few others are likely to be developed soon, since the least costly sites have already been used, and public opinion runs against development of many areas owing to the substantial opportunity costs represented by environmental and recreational uses.

Against the backdrop of this increasingly competitive market for water, agriculture is both the greatest consumer of water and the most highly subsidized user among those consumers that pay for water. Agriculture currently accounts for over 80 percent of all water use in the state, and also receives water sold at prices below the true value of the resource. Although considerable opportunity exists for conservation by municipal users—whose consumption of water is substantially above basic need levels—political realities suggest that agriculture, rather than other sectors, is likely to give up some of its share. This outcome is most likely to be achieved through price mechanisms that will make increased efficiency of water use on farms an economic necessity.

The ability of agriculture to maintain present acreages and yields despite a decreasing share in the water supply is questionable. Scientists disagree on the extent to which conservation measures will compensate for reduced supplies. Some experts are concerned that if severe shortages arise in the future because of factors such as continued population growth or climate change, agriculture will already have exhausted its conservation potential and will have no margin within which to buffer the effects of shortage. They argue that because both surface-storage and groundwater-recharge systems take years to plan, society will suffer the effects of insufficient water supplies long after the political will to develop more storage capacity has emerged. Other scientists point out that although much effort has already been devoted to finding water conservation methodologies, room for increased efficiency exists. These scientists estimate that growers in some areas could increase the efficiency of their water use by 20 percent without significant yield reductions.

Effects on Crops of Reduced and Late Rainfall

Importance of Water.   Anyone who has grown a houseplant knows that an adequate supply of water is essential to its growth and survival. Inadequate or poorly timed application of water can result in retarded growth, early or late flowering, reduced quality and quantity of desirable plant parts, and death. A plant that is suffering from water stress is more likely to suffer from other problems as well. It will be more vulnerable to attack from insects and disease organisms and will lose ground in the competition against drought-tolerant weeds.

Because water is essential to all biological processes, there is little room to substitute other inputs to compensate for shortfalls. Although substantial effort has been devoted to breeding crops for drought tolerance, successes have been limited. The term *drought resistant* has usually meant that the crop can extract water that other species can't, or that the crop survives water stress better than other crops. Because of the fragile nature of this resistance, existing improvements could be inadequate to protect a crop from the wide fluctuations in rainfall and moisture conditions possible under global climate change.

Crop water demand is influenced primarily by the same environmental factors that may change because of the greenhouse effect. As previously noted, increased atmospheric concentrations can decrease demand for water by increasing crop water-use efficiency; but higher temperatures work in the opposite direction to increase water demand in support of higher evapotranspiration rates. The latter effect is likely to exceed the benefits of the former in many crops. Alfalfa is a good example of a crop whose water requirement is temperature-sensitive. The crop requires 5.1 acre feet per year when grown in Kern County, but 6.5 in Imperial County. This change is due at least partly to a temperature differential of about three degrees (Celsius).

Range and Dryland Crops.   In California's Mediterranean climate, rainfall and peak water demand for summer and perennial crops are not synchronized under even the best natural conditions. Grass and wildland habitats exist in precarious equilibrium as species compete for limited resources. The sensitivity of resident annual grassland, oak-savanna, or oak-woodland/grass systems to water availability is illustrated by the case of the blue oak (*Quercus douglasii*). This tree, the dominant native low-elevation tree in the state, currently suffers from an inability to survive beyond the seedling stage. Researchers believe that this failure to thrive stems from actions of the early Spanish settlers, who introduced annual grasses that were able to outcompete the native perennial grasses. These annuals use more soil water from the region mined by oak seedlings than did the native perennials, with the result that the only oaks standing today are those that germinated during periods of two or three consecutive wet years. The last such period occurred about sixty years ago. A drier environment caused by global warming could conceivably bring about the elimination of the blue oak in California.

The effects of decreased precipitation on rainfed winter annuals grown in California, such as oats, oat hay, barley, or wheat, will likely be different from those experienced by grasslands, owing to the flexibility farmers have in management of these crops. Elimination of species is unlikely, since plantings of winter wheat and specialty crops will follow shifting rainfall patterns. In marginal rainfall areas, crops with low water requirements will be preferred, or, where slopes allow, irrigation will be instituted on an as-needed basis. This type of irrigation is already being practiced by some wheat farmers, who cut small furrows in their fields to allow irrigation when required. Other cropping systems that increase the flexibility of response to drought, and management practices such as water harvesting and field-level storage, will assume new importance.

Although irrigation will raise the cost of growing some traditionally rainfed winter crops, water used early to midway through the rainy season may not represent a net loss to total stored supplies. Winter irrigations may simply reduce some of the excess water that would be lost through high streamflows and flooding.

Timing of precipitation will be just as important to range and dryland crops as total quantity of water. A delay in the onset of the rainy season could result in serious decreases in autumn forage production for domestic livestock and wildlife grazers; annual grassland range plants typically germinate in autumn, grow rapidly until low temperatures nearly stop aboveground productivity during winter, and grow very rapidly in late winter and early spring until soil water is depleted. Other forage grasses would be impacted as well by a late rainy season: some perennials regrow in autumn regardless of fall rains; these species would experience plant mortality due to extensive grazing and low moisture.

Rainfed winter annuals grown without the option to irrigate could be at risk from the late arrival of the rainy season should the temperature requirements of the crop no longer coincide with the availability of water. Wheat grown in the Central Valley, for instance, requires a certain period of cold temperatures to produce optimally. If rains came too late in fall, dryland wheat would ripen late in spring after temperatures had become too high.

Irrigated Crops. Decreased precipitation is likely to affect crop production on irrigated lands mainly through changes in pricing and distribution systems rather than directly as for dryland agriculture. Different results will occur depending on what means society uses to allocate scarce water resources. If strategies (already being employed in some areas when water supplies are low) to reduce the length of the irrigation season are pursued, researchers might breed crops that develop in a shorter time period than those presently under cultivation. If water is allocated through a price mechanism, water costs to agriculture could increase dramatically, and the mix of crops planted in the state could change. Which crops would dominate the mix would depend as much on the elasticities of demand for the crops under consideration as on crop water-use efficiencies.

Regardless of the mechanism chosen to ration water among users, agriculture will need to follow water-conserving strategies. Many such practices are already followed. Pressurized-delivery systems are used to assure even application of water to each crop plant in a field, with the added benefit of reducing competition from weeds and evaporative losses. Some crop rotations include a fallow season during which the soil is stubble-mulched, then disked shallowly to preserve one season's soil water for the next year's crop. A concept known as "deficit irrigation" is thought to have potential for economizing on water without compromising yields. In perennial fruit crops, which have a large water requirement—three to four acre/feet per year—applied water can be reduced by up to 20 percent without yield loss if drought stress occurs at the right points during fruit development. Unfortunately, the long-term effects of this deficit irrigation are not clear, and greater control over the timing of irrigation would be needed by growers in order to implement the strategy.

Effects on Crops of Increased Rainfall and Variability
Increased rainfall should benefit California agriculture in the long run, although more flooding and erosion might occur in the near term, and costs due to variability and unpredictability of rainfall would be exacted as growers attempted to adapt. Depending on the temporal and geographic distribution of the increased precipitation, the types of changes induced by greater rainfall would be similar to those resulting

from decreased rainfall: preferred cropping patterns and management systems would change. Acreages of high water-requirement crops like rice and alfalfa might increase, and crops like safflower that can alleviate drainage problems might be added to rotation schedules. Safflower is currently planted in rotations where perched water tables exist, since its deep tap root can extract water from depths of up to six or eight feet, thereby lowering the water level for the following irrigated crop.

If rains arrive at the wrong time in crop development, increased rainfall could have negative effects. Periodic rainfall in summer could increase blight in walnut and tomato, brown rot in stone fruits, rust on beans, and powdery mildew on melons, to mention just a few of the potential pathogen problems. Unpredictable storms could ruin tree crops by washing away pollen during flowering. The cost of producing raisins could increase dramatically if rains were to necessitate mechanical drying of grapes. To the extent that rainfall is predictable and consistent, problems can be minimized by adaptive cultural practices. However, if future precipitation is characterized by great variability, agricultural losses could be substantial.

Effects on Animals

Much livestock feed is composed of crops that have high water requirements. If water prices increase, alfalfa and corn silage production could become uneconomic, making these important feeds less available for California livestock. Because conversion of plant matter to animal flesh is a relatively inefficient process, much water is used in the production of one pound of livestock.

Adding further to the high water requirement of livestock, current cooling and cleaning technologies are water-intensive. Should water prices increase substantially because of global warming, potential exists to lose the state's dairy, swine, cattle feeding, and poultry industries. If water prices do not rise, but rainfall and humidity increase, livestock could suffer from higher rates of disease. Hot, humid conditions, for instance, are known to increase the incidence of mastitis—the most costly of all diseases in dairy cows. If increased precipitation also resulted in standing water, populations of culicoides, mosquitoes, and snails would multiply. Snails, like culicoides and mosquitoes, are important in the life cycle of animal diseases, playing intermediate host to liver flukes of cattle and sheep.

Fish may be more sensitive to changes in precipitation than livestock. Increased flooding from untimely or increased rainfall could raise sediment and nutrient loads in streams and ponds. Decreased rainfall could have several unfavorable consequences. If water prices rise, fisheries will need to find alternative methods for flushing excessive nutrient loads out of ponds. At the same time, lower streamflows might result in increased concentrations of pollutants in stream waters, with attendant risks to fish and animals farther up the food chain.

Water Quality Issues

Fish will not be the only agricultural commodity to suffer from water pollution. Pollution in the form of salinity and buildup of toxic elements already imperil agricultural production in some parts of the state. Livestock and humans may be at risk from pesticides and fertilizers that have been detected in surface runoff and groundwater. If precipitation should decrease (or even remain the same with increased evapotranspiration due to global warming), water-quality issues could become even more important than they are at present.

Soil Salinity and Buildup of Toxic Elements. About three million acres of irrigated cropland in California are presently affected by saline soils, and an additional 1.6 million acres of nonsaline soils are affected by the salts in irrigation water. The west side and southern end of the San Joaquin Valley are most affected, but up to half of all irrigated acreage in the state is judged susceptible to loss of production due to salinity and sodic problems.

Selenium, arsenic, boron, mercury, and other naturally occurring trace elements that are toxic to plants or animals are natural constituents of some Central Valley soils. When these soils are irrigated under conditions of poor drainage, the undesired elements can build up to toxic levels. Such an event was discovered to have occurred in 1983 at Kesterson Reservoir, where selenium from irrigation drainage waters accumulated in concentrations that killed or injured fish and waterfowl. Selenium toxicosis has also occurred in livestock in the past, and management practices have been changed to control the problem.

Pesticide and Fertilizer Contamination. Pesticides and fertilizers have been found in runoff and groundwaters downstream from agricultural areas. A recent State Water Resources Control Board study cites over 2,900 verified incidents of groundwater contamination by pesticides in twenty-eight counties. Over fifty distinct pesticide compounds were identified, although only one—the long-banned nematocide DBCP (1,2 dibromo-3-chloro propane)—was widely distributed. In the Colusa Basin Drain, rice pesticides have been implicated in fish kills, while municipal water supplies in Sacramento have had periodic taste and smell problems stemming from use of molinate, a rice herbicide.

Groundwater monitoring has shown an increasing number of incidents of nitrate contamination during the past several decades, attributed to growth in both population and food production. Between 1965 and the early 1980s, use of fertilizer nitrogen increased by 67 percent.

This rate of use has since leveled off, but contamination by nitrate remains a cause for concern owing to linkages between nitrate consumption and "blue baby" syndrome (methemoglobinemia).

Effects of Global Warming on Water Quality.   If global warming results in a decreased supply of water for agriculture, higher evapotranspiration rates combined with reduction in water available to leach salts from soils could increase the rate of accumulation of salts and toxic elements in soils. Increased temperatures may also boost the rate of volatilization of pesticides from crops, meaning that more sprayings will be required. Potential for an even greater impact on water quality due to pesticide and fertilizer contamination may exist if climate change results in increased or more erratic rainfall—which could increase leaching and surface removal of applied chemicals, resulting in greater contamination of runoff and groundwater.

Higher sea levels created by melting ice and thermal expansion of ocean waters may degrade coastal and delta zones that currently produce high-value crops. Increased freshwater flows would be required to control saltwater intrusion resulting from rising sea levels, and this would place great pressure on already weakened levees. Without enhanced freshwater flows, the quality of water supplied through the Delta would deteriorate in adjacent lands and in urban and agricultural waters transferred to the southern sections of the state. Currently the Delta transfers 47 percent of the total runoff and provides 45 percent of the state's drinking water and 40 percent of the agricultural water supplies. Loss of the Delta system would be an economic catastrophe. It has been estimated it would cost one billion dollars to renovate the levee system just to serve current needs. The cost of levee repairs does not include the potential economic losses that would result from reduced agricultural productivity and direct loss of some agricultural land in the Delta if the levees are not repaired. Anticipation of a significant rise in sea level would require even higher and stronger levees, escalating those costs.

EFFECTS OF INCREASED AIR POLLUTION AND UV RADIATION

The Outlook for Air Quality

California fails to meet EPA air quality standards more often than any other state in the nation, despite stringent state regulations on both mobile and stationary emitters. This nonattainment of standards stems from several factors, including the state's geography, climate, and large population. Although none of these factors can be directly controlled by legislation, mandates for the use of improved technologies such as catalytic converters on vehicles have been successful in improving the quality of some air basins. Other areas have failed to improve, despite reductions in reactive organic gases over the last ten years. Within the Central

Valley, Fresno and Kings counties currently experience worse ozone levels than New York, Houston, Philadelphia, and Chicago. Unfortunately, the future outlook for air quality in the valley is bleak. Demographic trends alone will virtually guarantee growth in pollutant levels, as population growth rates in the Central Valley currently exceed those for the state as a whole. In addition, damage due to both the direct effects of UV_b radiation and the indirect effects of radiation on photochemical reactions is likely to increase as chlorinated fluorocarbons deplete stratospheric ozone.

Global climate change will tend to exacerbate the effects of the various forces that deteriorate air quality. The rate of emissions and the chemical transformations of those emissions to harmful compounds are likely to increase. Emissions of both anthropogenic and natural origin will grow in response to higher temperatures due to the acceleration of metabolic processes in living organisms, the increased volatilization of some pesticides, solvents, and fuels, and the increased energy generation needed for cooling purposes. The rate of transformation of emitted chemicals to harmful compounds will grow, since increased temperature and UV radiation drive these reactions also. Finally, the percentage of the year during which poor air quality prevails could lengthen because of climate change if warmer temperatures create stronger surface inversions in winter, or if the rainy season commences later in the fall.

Magnitude of the Problem

Ozone is the primary atmospheric pollutant causing injury to vegetation in California. Estimates of crop losses in the Central Valley due to ambient ozone levels range from over 20 percent for beans, melons, and grapes to 9–15 percent for alfalfa, alfalfa seed, cotton, lemons, oranges, and potatoes. Together with other photochemical oxidants, ozone is estimated to cause economic losses in the range of several hundred million dollars a year in California. Losses due to wet and dry deposition of acidic substances in some of California's airsheds may add to the total, but few resources have been devoted to quantifying these effects.

Physiological Response of Crops

Low levels of exposure to phytotoxicants over long time periods reduce yields more than do quick, high level concentrations. Evidence suggests that deciduous fruit trees are particularly susceptible to long-term exposure to atmospheric toxicants. Recent screening of nine prominent tree crops indicated that photosynthesis and growth rates in several of the species tested are substantially reduced when the plants are grown at ozone concentrations equal to twice the ambient levels existing at the University of California's field station located near Fresno. Among those tested, the crops that demonstrated the greatest sensitivity were almond, apricot, plum, prune, pear, and apple. Peach, nectarine, and cherry showed less susceptibility. Some growers have expressed concern over a perceived potential for air pollutants to create surface blemishes on peaches, plums, and nectarines; however, little research has been done to substantiate this effect.

Detrimental effects of air pollution on plants can have the same consequences for crop survival and quality that are induced by temperature and moisture stresses. Stressed plants and animals may become more susceptible to attack from pests and pathogens or less able to compete with undesirable species. For example, increased UV radiation has been shown to favor growth of wild oat over wheat in cultivated fields. Additionally, damage to cells from oxidants and UV radiation may increase requirements for nutrients, while water stress may raise plant susceptibility to ozone damage.

Physiological Effects on Animals

Skin cancers and eye and respiratory problems in livestock will increase as air quality deteriorates. Increased UV radiation is the principal cause of eye cancer in cattle and skin cancer in light-colored cattle and sheep. Ozone is an irritant to the respiratory system, which predisposes animals to bronchitis and pneumonias. Nitric acid deposition can cause eutrophication in fisheries that draw water from affected watersheds, such as in the central Sierra Nevada. Fish living in high-elevation, poorly buffered lakes could suffer the direct effects of increased acidity.

Potential Responses

Because differences in susceptibility to damage from air pollution exist among species, there may be potential to improve genetically the tolerance of plant varieties to various air pollutants. However, reactive chemical species such as ozone, photochemical oxidants, and UV_b radiation—which causes formation of free radicals—can affect most biochemical reactions essential to the survival and reproduction of organisms. Creation of true resistance to the effects of these compounds would require changes in the basic chemistry of life. Therefore, pollution prevention will remain the best strategy for avoiding toxic effects due to these compounds.

Agriculture itself may have potential to make a small contribution toward pollution prevention. While most plants emit almost no reactive gases, other species contribute to air-quality problems by emitting hydrocarbons. Isoprene and monoterpenes emitted by plants are about three times more reactive than the average anthropogenic mix in urban areas. It has been suggested that some urban areas would not meet national air-quality standards even if all anthropogenic hydrocarbon sources were removed, because of the gaseous emissions from local vegetation alone. Because plants vary greatly in their emissions, the problem of hydrocarbon contribution can be alleviated by an appropriate choice of plant species. Research is ongoing to test the emissions of the state's major crop species.

INTERACTIONS WITHIN AND BETWEEN SYSTEMS

California agriculture has evolved in response to changes in the physical, biological, economic, and social systems that constitute its total environment. A stimulus for change in any one of these systems can have repercussions within the other systems and result in changes in the technologies and management decisions appropriate for agriculture that would not be predicted on the basis of knowledge of the initial stimulus alone. For this reason, an assessment of the effects of global warming across systems is both essential and difficult.

Biological and Physical Systems

Interrelationships within Systems. Scientists who attempt to model ecosystems find that their models most closely approximate reality when they are based on the reactions of individuals rather than on generalized population parameters. This is because the number and diversity of organisms that evolve in a given situation depend on the initial availability to each individual of nutrients, water, light, air, and other resources. The proximity and nature of other organisms in the immediate environment further influence the growth of the individual.

A simulation experiment designed to study the response of boreal forest in Minnesota to a warmer, drier climate demonstrates the complexity of interactions that determine a system. The study showed that the moisture storage capacity of two different soils would have a critical effect on interactions among tree species, leading to a divergence in species composition, soil-nitrogen availability, and forest productivity. The implications of these findings for scientists attempting to reduce the transitional costs due to global warming are clear. Detailed information will be required on factors such as crop requirements for heat, moisture, and clean air; effects of beneficial insects, pest predators, disease vectors, and alternate hosts; processes of soil microorganisms that affect organic matter, soil fertility, quality and nutritional content of plant matter, and carbon and energy cycling. These are but a small subset of the factors that influence crop development; together with other factors they will determine the nature and effectiveness of attempts to locate new regions of adaptation for crops in a warmer world.

Interactions across Systems. Today's agriculture depends on much more than the physical and biological systems that control plant or animal growth. California agriculture has adapted and readapted to changes in the social and economic environment over the past seventy-five years as the population has increased and the prices of labor and land have risen relative to other inputs. Introductions of massive-scale irrigation, farm machinery, pesticides, herbicides, and chemical fertilizers have resulted in production of quality food and fiber sold at generally affordable prices.

Unfortunately, these affordable prices have not reflected the costs to the environment of the contaminated air, land, and water resources that result from some modern practices. With growing public awareness of the importance of unanticipated side effects, constraints on the use of current technologies are increasing. Loss of current technologies due to regulation will simultaneously drive up food production costs and reduce the options available to agriculture for adapting to future change. The need to protect the environment could therefore decrease the resiliency of agriculture to the effects of global warming.

Economic and Social Systems

Infrastructure.   Cropping patterns in California are characterized by regional preferences for certain dominant cropping systems. These preferences may be traced in part to climatic factors, but are also attributable to the existing infrastructure that has built up over time in support of the dominant crops and animals. Infrastructure means access to cooperatives, expert advice, farm equipment and repair facilities, and processing plants and markets. These resources all contribute to the viability and efficiency of local agriculture. Without them, information, marketing, and transportation costs would rise dramatically.

An expected result of global warming is that cropping patterns will change. Although changes may result in net gains for agriculture in the long term (if and when the climate has stabilized at a new equilibrium state), the inflexibility of local infrastructures could raise production costs in the interim.

Policy Environment.   Public policies create an environment conducive to certain cropping systems and also promote development of infrastructure. Local zoning, regional water pricing, state regulation of pollutant discharges, and national price supports and tariffs are some of the many policies formulated by governments that affect farmers. Wise and timely policies can reduce uncertainty and risk in farming, while less advantageous ones create inefficiencies. The effects on agriculture of policy responses to global warming will be correspondingly diverse. Although discussion of these policies is beyond the scope of this chapter, some general observations can be made.

Present policies exist in response to needs that have been identified in the past. Climate change will necessitate enactment of a new set of policies. However, there is likely to be a lag period while consensus develops regarding the direction that new policies should follow. Existing policies may impede the ability of farmers to adjust to climatic change in the near term. Rigidity in policy structures already prevents some advantageous changes in cropping systems under the present, relatively stable climate—as illustrated by existing commodity programs, which require maintenance of crop acreage allotments as a basis for price support payments, thus discouraging investment in alternative cropping systems. Climate change itself may increase the rigidity of policy structures if shifts in comparative advantage threaten nations' abilities to continue economic production of necessary foodstuffs. Events less profound than global warming have resulted in worldwide tariff wars. Imposition of trade barriers reduces world GNP, and could potentially cause changes in global desire for certain commodities due to the effects of reduced income on demand curves.

Because climate change will create instability in economic systems, a shifting of the relative power held by political interest groups is likely to occur. Given the increasing number of Californians who are employed in occupations other than farming, policies may be enacted that are unfavorable to the agricultural sector. Research and public education will be essential to assure that new policies reflect society's true, informed preferences.

Markets. While the uncertainty in climate models makes it difficult to predict the changes that will occur in global markets, the likelihood that some change will take place is high. California is able to maintain its preeminent position in national and global agricultural markets primarily because of the high quality of the specialty crops that it produces. At present, 95 percent of U.S. processing tomatoes are grown in California because the stable climate, which includes dry summers, allows production of unblemished fruits. If summer rains cease elsewhere in the country, those states with irrigation potential could begin to compete for tomato markets.

The cost of agricultural inputs will be an additional factor determining California's competitiveness. For example, increased water prices might result in the elimination of rice production, since rice has a high water requirement and is sold on the international market where it must compete with rice grown with inexpensive water. Rising energy costs could reduce the profitability of crops such as cotton, sugar beets, and alfalfa owing to higher water pumping and tillage costs.

Management Decisions. Beset by a changing environment, agricultural managers are likely to pursue strategies that reduce their risk. At the cropping systems level, farmers may make trade-offs between dryland versus irrigated cropping systems and cropping versus range systems. If water pricing and allocation changes, growers might decide to avoid risk altogether, and forgo less profitable irrigations. To gain the flexibility of deciding which crop to plant each year, farmers might prefer annuals over perennials, since production of perennials would entail added risks due to the danger that the climate could change during the productive life of the crop. Controlled-environment, glass-house production of vegetables might increase near population centers to fill the demand for premium-quality produce when field-grown crops suffer climate-related damage.

At the farming systems level, more farmers might choose to rent their cropland rather than buy. The optimum scale of farming might change. Integrated corporations might diversify their production units even more than at present by growing several crops in diverse locations and controlling transportation and processing facilities.

INFORMATION NEEDS

In order to formulate rational responses to climate change, much information will be needed regarding the systems that influence agriculture. Some of this information currently exists, but the rest can be acquired only through additional research. The following discussion outlines priority areas for information gathering and research that may help to minimize the disruptive effects of a changing climate.

Existing Databases and Information

The design of experiments in the agricultural sciences has traditionally yielded information that is narrowly defined and specific to the exact factors tested. Volumes of data have thus been generated, giving a good picture of some processes but almost no understanding of others. Because climate change threatens to entail extensive alterations in agricultural systems, it is essential that existing information be compiled in a systematic and comprehensive way. Some successes at such endeavors already exist. CIMIS, the California Irrigation Management Information System, provides information on local evaporative rates, as well as weather predictions that help farmers decide when to irrigate. Researchers in the University of California's IPM (Integrated Pest Management) program have created a database with information on pest management in several important commodities, which is available to the general public. The university-sponsored Sustainable Agriculture Research and Education Program is undertaking the mammoth task of computerizing a database on publications that provide information on sustainable practices. This database will be available to user groups interested in applying sustainable techniques to their systems.

Computer modeling is one means of using data that allows tentative predictions of the behavior of systems in response to specific changes in the environment. Models that predict yield under various temperature and moisture conditions exist for wheat, cotton, alfalfa, potatoes, sugar beets, and rice, and ecosystem models exist for some grasslands. Similar models would be useful for other dominant crop species as well. Models predicting feed availability and price could prove critical to the dairy and beef cattle, poultry, swine, and small ruminant industries. Livestock growers could likewise benefit from the output of models that estimate the range and distribution of pest species under altered climates.

Research Priorities

Research will be needed to fill in the gaps in existing knowledge and to focus specifically on the changes expected from global warming. The information gained will be useful in at least three ways: it will help solve today's problems; it will foster development of adaptive technologies; and it will set the stage for extension and public education aimed at better future management and policy decisions.

To supply crucial information for policymakers and the general public,

Research on economic factors should:

- Estimate the economic impacts on various crops and animal industries of price changes for agricultural inputs such as water and energy. Although many such estimates now exist, they should be compiled and missing commodities added.
- Project changes in California's comparative advantage under plausible climate-change scenarios.
- Provide cost-benefit analyses of urban encroachment on agricultural land and effects of resultant air pollution on crop yields.
- Lay out alternatives for resolving the salinity and toxicity problems confronting the San Joaquin Valley. Research results should translate the problem into economic terms that state the bottom line to policymakers.

Research on sociological dimensions should:

- Increase understanding of the decision criteria employed by farm managers when choosing cropping systems and production technologies.

Research to increase scientific knowledge should:

- Improve understanding and communication of the risks to animals (including humans) from exposure to agricultural chemicals in water, air, and foods.
- Assess the trade-offs between yields and the preservation of the environment through low-input agricultural systems that might be mandated by environmental legislation.

To provide information useful in solving technological problems, research will be needed that falls into three categories: 1) that research which is needed whether or not global warming occurs, 2) that which helps define the parameters of climate change, and 3) that which will provide information essential for the mitigation of the impacts of warming on agricultural productivity and sustainability.

Research needed in any case should:

- Increase understanding of flexibility within cropping systems, especially through study of crop ecology. Create new systems to optimize production under additional constraints.
- Expand awareness of the dynamics of carbon in soils, marshes, bogs, anoxic sediments, and peats.
- Increase understanding of watershed dynamics, development, and management.
- Develop crop and pest management strategies that emphasize cultural practices rather than chemical controls.
- Improve efficiency of water distribution through comparative studies of available techniques.
- Increase efficiency of water application by farmers, and use by plants and animals.
- Define the effects of air pollutants on crops and livestock, and screen for genetic variability in susceptibility to pollutants. Determination should be made of the consequences of longer smog seasons resulting from drier autumns.
- Increase knowledge of DNA repair mechanisms useful in attenuating the biological consequences of UV radiation.
- Increase the efficiency of energy conversion during biofuel production. Could include study of cellulose biosynthesis and the physiology and biochemistry of lignin degradation, as well as quantification of total volatilization and emissions from biofuel use.
- Foster development of energy- and water-efficient degradation of domestic organic wastes.

Research that defines the parameters of climate change should:

- Improve the resolution and accuracy of outputs from climate models. More information on the size and direction of weather changes expected in specific regions is needed to guide research in other disciplines.
- Determine the effects of temperature and enhanced UV radiation on formation of both photochemical smog in the Central Valley and hydrocarbon and NOx precursor emissions due to increased energy use.

Research to reduce the impacts of climate change should:

- Provide complete data on crops and livestock regarding temperature and moisture sensitivities.
- Define the chilling requirements of deciduous fruit and nut crops grown in the valleys of the state under both today's climate and hypothetical future climates.

- Increase genetic tolerance to heat stress in crop species, and provide efficient cooling mechanisms for use in livestock production.
- Increase understanding of the interactions that occur because of a combination of temperature and water stress, as well as increased CO_2 concentrations and photoperiod requirements. Nutritional quality of crops and livestock, total biomass production, susceptibility to pests and pathogens, and competition from weeds should be considered.
- Promote understanding of the effects of modified atmospheres on crops and livestock and air pollution, as well as carbon cycling. Construction of controlled atmosphere facilities will be needed to expedite this research.
- Evaluate the effect of elevated CO_2 levels on crop productivity, water-use efficiency, and crop development and partitioning of substrates. Specific information on affected physiological and biochemical processes is needed for all biological systems.
- Quantify the potential of agroforestry to store CO_2 as plant biomass.
- Quantify the hydrocarbon emissions from plants, and the effects of increased temperature on such emissions, with special focus on predominant tree and crop species.
- Provide a greater understanding of soil processes under increased temperatures. Effects on soil salinity, organic matter, structure, and nutrient availability could be important.

CONCLUSION

This paper has highlighted some of the impacts that a warmer climate may have on agriculture in California. Because of the state's diverse geomorphology it is difficult to predict what crops will grow in which locations under future climate regimes. However, the potential interactions between warmer temperatures, higher CO_2 concentrations, and the factors that affect plant and animal growth may have major consequences for the competitive position of the state's agriculture. Forward-thinking research and public policies are required to assure that responses to climate change will optimize production systems under future constraints.

REFERENCES

California Energy Commission. 1989. *The impacts of global warming on California: Interim report.* Sacramento.

Coppock, Ray. Undated. *Resources at risk: Agricultural drainage in the San Joaquin Valley.* Davis, Calif.: Agricultural Issues Center, University of California, Davis.

Demment, Montague, K. Cassman, W. Chancellor, D. Chaney, E. Learn,

R. Loomis, D. Muns, D. Nielsen, J. Seiber, and F. Zalom. 1989. Farming systems in California: Diversity to compete in a changing world. Unpublished.

Department of Water Resources. 1987. *California water: Looking to the future.* Bulletin #160-8. Sacramento.

Engelbert, E. A., and A. F. Scheuring, eds. 1982. *Competition for California water: Alternative resolutions.* Berkeley, Los Angeles, London: University of California Press.

Huston, M., D. DeAngelis, and W. Post. 1988. New computer models unify ecological theory. *BioSci.* vol. 38, no. 10.

Menke, John. 1989. Management controls on productivity. In *Grassland structure and function: California annual grassland,* ed. L. F. Huenneke and H. A. Mooney, 173–199. Dordrecht, Netherlands: Kluwer Academic Publishers.

Pastor, J., and W. Post. 1986. Influence of climate, soil moisture, and succession on forest carbon and nitrogen cycles. *Biogeochemistry* 2:3–27.

Pendleton, D., and G. VanDyne. 1982. Research issues in grazing lands under changing climate. In *Environmental and societal consequences of a possible CO₂-induced climate change,* vol. 2, pt. 16. Contract #DE-AS01-79EV100. Washington, D.C.: United States Department of Energy.

Singer, Michael, W. Wood, Jr., and C. Lynn. 1990. Farmland in California: A changing resource. In *California agriculture during the next two decades: Choices and constraints.* Davis, Calif.: Agricultural Issues Center, University of California, Davis.

Vaux, Henry, and R. Woods. 1990. Water for California agriculture: The current status and future challenges. In *California agriculture during the next two decades: Choices and constraints.* Davis, Calif.: Agricultural Issues Center, University of California, Davis.

Winer, Arthur. 1989. *Hydrocarbon emissions from vegetation found in California's Central Valley.* California Air Resources Board Contract # A732-155. Riverside, Calif.: Statewide Air Pollution Research Center, University of California, Riverside.

———. Undated. *Potential air quality impacts on agriculture resulting from global atmospheric modification.* Riverside, Calif.: Statewide Air Pollution Research Center, University of California, Riverside.

Winer, Arthur, D. Olszyk, and R. Howitt. 1990. Air quality: An endangered resource for agriculture. In *California agriculture during the next two decades: Choices and constraints.* Davis, Calif.: Agricultural Issues Center, University of California, Davis.

SEVEN

Global Climate Change
and California's Natural Ecosystems

Daniel B. Botkin, Robert A. Nisbet, Susan Bicknell,
Charles Woodhouse, Barbara Bentley, and Wayne Ferren

Nationally, there are many reasons why considerable attention should be focused on the impact of climatic change on California: it has great biological diversity, many endangered and threatened species, a number of important national parks and national forests, and its forests and wildlife are a familiar symbol worldwide for biological conservation (see fig. 1 for a representation of present California ecosystems).

If projections of global climate models are correct, the natural ecosystems of California might undergo major changes during the next century. Such changes might include large economic losses in timber, fisheries, and recreation; major changes in our national and state parks and forests and in our nature preserves and conservation areas; increase in extinction of endangered species; loss of large areas of existing habitats; and development of new habitats whose location and areal extent can only be surmised. Many areas currently set aside for the conservation of specific ecosystems might no longer be suitable to them. Yet, in spite of the potential seriousness of these problems, which could dwarf all other environmental changes, California is at present in a poor situation to project what the effects of global change on its natural ecosystems might be.

Most considerations of the ecological impacts of global warming have focused on long-term consequences, but we must be alert to transitional effects that could last decades, during which natural ecosystems would undergo large-scale disruptions. We know the most about and are in the best position to make projections for forest vegetation and coastal oceans and wetlands, but even for these areas projections remain informal and qualitative. For example, while it may be true that one forest type will be replaced over the long run by another, existing forests might decline before warmer- and drier-adapted species are able to immigrate and

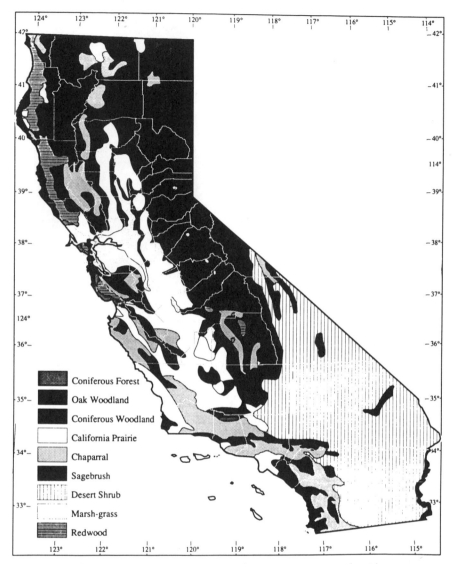

Figure 1. Natural vegetation zones in California. (Map courtesy of Bern Kreissman, *California: An Environmental Atlas and Guide* [Davis: Bear Klaw Press, 1991])

become established, and there may well be a period of tree die-offs and a temporary loss of habitat for certain wildlife species (Davis 1989). Species with restricted ranges and specialized environmental requirements, such as the giant sequoia, might be caught in ecological "islands" without sufficient contiguous refuge areas where seeds could migrate and germinate. The very long-lived sequoia and redwood might persist through this transition, but most woody vegetation would be severely disrupted. Increased temperature and decreased soil moisture might lead to an increase in fire frequency, accelerating vegetation changes and in the long run leading to a decrease in forest biomass. If rapid climate change were to continue past the twenty-first century, forests as we know them might no longer be sustainable, since a tree seed that might germinate in a suitable climate in one decade might reach maturity in a climate no longer suitable to its growth or the germination and survival of its seeds. If such conditions were to occur, only the shortest- and longest-lived tree species might persist. Trees common over much of California's landscape today might be no longer viable.

While it is clear that the natural ecosystems of California will undergo alterations, our ability to make quantitative projections about the impacts of climate change lags behind that for the eastern United States. Projections made for the forests of eastern North America suggest that the anticipated climate changes will lead to dramatic and surprisingly rapid ecological alterations, including large-scale changes in the distribution of major species of vegetation, with some effects observable by the turn of the century or soon after (Botkin and Nisbet 1990; Smith and Tirpak 1989). In part, our inability to make similar predictions with confidence is due to California's multitude of physiographic regions and its complex flora and fauna; but it is also due to the slower pace of ecosystem research and the development of adequate ecological models for California ecosystems. *At this time we are lacking in our ability to make projections necessary to the development of wise policy for the management of many natural ecosystems, endangered species, and many commercially important renewable biological resources.* The best we can do is to point out likely qualitative responses of natural ecosystems to global warming. The challenge that faces us is thus threefold:

1) to set down, according to the best current state of knowledge, what qualitative ecological changes are likely;
2) to set forth a research program to be carried out during the next five years which will make possible quantitative projections;
3) and to suggest what solutions there might be to sustain the natural ecosystems of the state, protect threatened and endangered species, maintain the economically important timber and fisheries resources, and conserve the scenic resources of marine, coastal, and upland ecosystems.

MARINE ECOSYSTEMS

What We Know about Marine Ecosystems

Our general knowledge of oceanic ecosystems can be applied regionally to develop an understanding of the processes that control these systems and their response to environmental perturbations. The ocean is the world's largest reservoir for carbon. It exists there in the form of dissolved CO_2, carbonate and bicarbonate, $CaCO_3$ in sediments, and as dissolved organic carbon (DOC). The ocean is also a very large heat reservoir. To a very great extent the consequences of adding more CO_2 to the atmosphere depends on how the ocean responds to a change in global heat gradients and therefore the wind-driven current systems. These current systems regulate levels of primary productivity by recycling plant nutrients. It is this productivity that results in the vast areas covered by $CaCO_3$ sediment and the enormous pool of DOC.

Ocean Monitoring. The ocean along the California coast has been monitored on a regular basis since 1949 by the California Cooperative Oceanic Fisheries Investigation (CalCOFI). This program has been in place during two major and a number of lesser warming events termed El Niño Southern Oscillation (ENSO). Consequently an extensive database exists, on which marine scientists can draw to formulate educated speculations on the consequences of meteorologic and oceanographic warming trends in California. ENSO events are usually signaled by an increased atmospheric pressure difference in the southern Pacific, leading to an intensification of winds and subsurface temperature along the equator, propagating northward. Warming of water due to global warming may or may not trigger other ENSO-like changes.

Coastal Upwelling System. California's coastal upwelling system is well known, including its variability with respect to intensity or degree of vertical advection with time. Upwelling events occur as pulses, which are presently on the order of three to six days. The pulsing pattern is biologically important to nearshore species such as fish larvae. Upwelling itself is associated with deleterious effects of offshore transport and turbulent mixing which dilutes food concentrations.

Onshore Airflow. The cool upwelled surface waters in turn cool and stabilize onshore airflow, which creates a cool summer climate along the coast. The longshore wind which drives coastal upwelling is partly maintained by an atmospheric pressure gradient between the "thermal" low pressure cell that develops over California's heated interior and the higher pressure maintained over the cooler ocean.

Wind Stress. An increase in the equator-ward wind stress over the California Current has been observed since 1946. A similar increase has also been noted for other major upwelling systems. With a warming trend, wind intensity and upwelling intensity would increase. Consequently, one cannot simply expect sea-surface isotherms to increase because of effects of other physical forces such as wind stress. Furthermore, upwelling intensity and degree of turbulence are apparently linked to the recruitment of key food organisms such as pelagic fish (e.g., northern anchovy).

Horizontal Advection.   In addition to coastal upwelling, California's marine ecosystems are strongly influenced by horizontal advection (lateral flow) from the north. Both processes bring nutrients to the system, although upwelling is probably more important in contributing nutrients very near the shore and horizontal advection more important in the main body of the California Current. Both sources may be affected by global warming. A decrease in the California Current's strength appears to be characteristic during ENSO episodes (fig. 2), hence global warming might create a similar decrease. ENSO events can serve as one model of the consequences of warming in this current system (Chelton 1981; Chelton, Bernal, and McGowan 1982; Simpson 1984; McGowan 1985; Hickey 1979).

Anoxia.   A gradual change in the characteristics of marine ecosystems can have profound effects on productivity in benthic and demersal communities or communities in poorly mixed environments. As water temperature increases owing to global warming, oxygen solubility will decline. This decline in oxygen solubility will facilitate the development of anoxic conditions, particularly below the photic zone, where decomposition of organic detritus can entirely consume the dissolved oxygen. Under these conditions, toxic substances such as ammonia and sulfides are generated.

Possible Impacts of Climate Change

Global warming may result in El Niño–like conditions. CalCOFI has monitored the California Current before, during, and after both the 1958–59 and the 1983–84 ENSOs, and these studies therefore serve as an excellent model for predicting effects of global warming on marine ecosystems (fig. 2). The forty-year CalCOFI biological-physical data set gathered during these ENSOs has to date been only partially analyzed, but from it we can suggest some plausible scenarios (Chelton, Bernal, and McGowan 1982):

- a generalized but patchy warming of the mixed layer;
- a lessening in the mass transport of water from the north;
- a deepening of the thermocline;
- less horizontal and vertical mixing and turbulence;
- much less primary and secondary productivity;
- appearances of migrant species from the south, but only near-shore;
- a large rise in sea level;
- clearer water;
- storms from the west and southwest, probably wet;
- an eventual great reduction in all traditional commercial and sports fish populations, as was the case during the ocean warming associated with recent California El Niños.

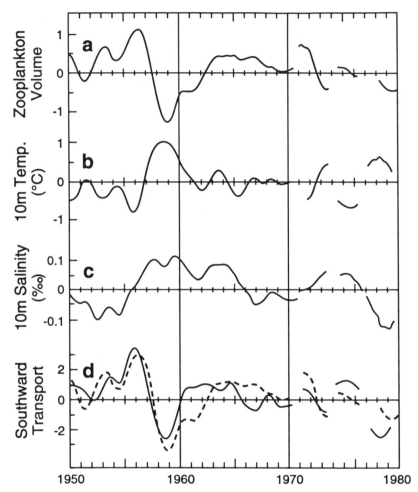

Figure 2. The low-frequency, nonseasonal anomalies frc:n the long-term means of ecosystem properties in the California Current. The data are space-averaged over a very large area. The axes are in standardized units vs. time. The dash line in panel (d) is the same as panel (a) zooplankton. This figure shows the very close relationship between the large variations in biomass and the variations in transport of cold, low salinity water from the north. The period 1958–59 was a strong El Niño period in this current system. (From Chelton et al. 1982)

There is a possibility of increased coastal upwelling due to the increased land-sea contrast in air temperature. The biological effect of this upwelling is partially dependent on the depth of the thermocline and nutricline. A slow California Current will probably be associated with a deep thermocline, and relatively little nutrient enhancement would be provided by upwelling under these conditions. An intensified California Current may be associated with a shallow thermocline, in which case coastal upwelling would be colder and nutrient-rich. The scenario of warmer terrestrial conditions in California does not necessarily equate to warmer marine conditions, especially during the spring and summer upwelling period. Furthermore, we can expect ENSO events to continue to be superimposed on the new system, with even warmer maximum ocean temperatures. The time of onset of the upwelling season begins earlier in the south than in the north. If the Pacific high moves northward, we can expect the onset of upwelling to occur earlier in the year.

Response of Fish. The response of various fish species to global climate change may be to change their productivity and migration patterns, but varying considerably among species.

Effects of Anoxia. In southern California, broad areas of undisturbed sediments suggest that areas of extreme anoxia, with oxygen concentrations of less than 0.1 mL/L, are expanding rapidly. If areas of anoxia expand because of increased temperature and declining oxygen solubility, certain economic impacts are to be expected. These include declining fisheries and general oceanic productivity, and declining recreational use of the coastal zone. In addition, an increase in hydrogen sulfide releases in coastal zones is likely.

Coastal-zone fisheries and other marine life will decline either by direct toxic effects of anoxia or by upwelling of toxic waters into the photic zone. The vast majority of the biomass of the ocean lives in the surface wind-mixed layer of the ocean, trophically dependent on photosynthesis. The development of anoxia, or rapid vertical advection of anoxic waters into the surface mixed layer, could create an ecologic crisis for various marine groups through a combination of 1) direct toxicity; 2) reduction or modification of nutrient or food supply; 3) chronic debilitation caused by contact with such deep waters; or 4) increased predation by more adaptive or less affected taxa. Developing anoxia affects a biotic community not only by lack of oxygen but also by changes in temperature, pH, and other constituents present in the water. The presence of chelating agents, antagonistic or synergistic elements, redox state, pH, and temperature modify the actual chemical configuration or activity experienced by the organism.

Along the California coast, the ocean is used as a dump for biological wastes, particularly from the Hyperion sewage plant in Los Angeles. The ability of the ocean to absorb biological wastes may rapidly diminish as the basins off southern California become increasingly anoxic. In extreme cases, hydrogen sulfide generation (such as that reported in a fjord in Oslo, Norway) causes blackening of houses and boats.

Effects on San Francisco Bay. Global warming would affect San Francisco Bay principally by causing sea-level rise. Higher sea level would flood present marshes and other tidal wetlands as well as push the entrapment zone upstream. Present levees would be breached, producing many shoal-water areas. Runoff from storm drains and from fields would lead to enhanced productivity in the expanded bay. At the same time, warming of surface waters would reduce oxygen content. Influx of waters from streams would decline as the upland soils became drier. Reduced water flow would tend to concentrate nutrients and organisms in the expanded bay. Reduced oxygen content and expanded shallow waters, rich in organisms, could lead to hypoxic conditions such as those experienced at present in similar environments in the Baltic and Adriatic. Global warming could lead to large areas of oxygen-deficient waters in areas now occupied by delta water. Indeed, global warming could result in reversion to conditions of the 1940s and early 1950s when bay tidelands were characterized by obnoxious odors and lack of marine life.

WETLAND ECOSYSTEMS

"Wetlands are lands transitional between terrestrial and aquatic systems where the water table is usually at or near the surface or the land is covered by shallow water" (U.S. Fish and Wildlife Service [Cowardin et al. 1979]). This classification requires the presence of at least one of three attributes: 1) hydrophytes; 2) hydric soil; or 3) nonsoil saturated by water or covered by shallow water at least periodically. Wetlands of the United States are classified (Cowardin et al. 1979) into five major systems (marine, estuarine, riverine, lacustrine, palustrine), all of which occur in California.

MARINE WETLANDS

Marine wetlands include intertidal wetlands and subtidal deepwater habitats of the open ocean, where they overlie the continental shelf and extend landward to the limit of tidal inundation (Cowardin et al. 1979). Along the coast of California, marine wetlands consist of wave-cut bedrock outcrops, sand and pebble beaches, and cobble deltas that generally are dominated by a rich association of algae, marine angiosperms, and invertebrates. Impacts to marine wetlands as a result of global warming will come largely from an anticipated rise in sea level. Accentuating the effect of sea-level rise is the simultaneous subsidence of some areas.

Potential impacts from sea-level rise will include:
- conversion of intertidal wetlands to deepwater habitats;
- net loss of habitat where habitat migration is not possible owing to land-use constraints;

- land-use conflicts exacerbated;
- increase of salinity in surviving habitats;
- potential failures of seawalls, inundating many urban, recreational, and wildlife-habitat areas.

Artificial barriers may exist or may be built to protect coastal real estate and could further limit the landward migration of marine wetlands. A rise in sea level of one-third meter could erode beaches from 67 to 133 meters in California (Titus 1989), but because many are not this wide, substantial loss of marine intertidal zones, wetland habitat, and recreational sites is likely in urban areas with seawalls. If the rise in sea level is one-half meter, Titus estimates one-third of coastal wetlands in the United States would be inundated.

ESTUARINE SYSTEMS

Estuarine systems include deepwater subtidal habitats and intertidal wetlands that generally are semienclosed by land but open at least sporadically to the ocean and receive at least occasional dilution from freshwater runoff (Cowardin et al. 1979). The upstream limit extends to where ocean-derived salts are diluted to 0.5 percent during the period of average annual low flow (Cowardin et al. 1979). In general, estuarine wetlands occur between the highest tide of the year and mean sea level. Prominent estuarine habitats are fairly well documented in terms of floral composition and distribution. Estuaries serve as habitats for many rare and protected species.

At least 75 percent of the estuarine wetlands of southern California are estimated to have been destroyed in the last century (Zedler 1982), and most of the remaining ones are either degraded, fragmented, or isolated remnants of historically larger wetland systems. Landward migration of wetlands as a result of sea-level rise will be constrained by abrupt topography of the coastline and artificial barriers constructed to protect agricultural lands and urban areas.

Possible Impacts of Climate Change

The EPA has estimated that from 40 to 73 percent of the wetlands in the United States could be lost by the year 2100 but that the potential formation of new wetlands might reduce this loss to from 22 to 56 percent (Armentano, Park, and Cloonan 1988). Other estimates include a 30 to 70 percent loss with a one-meter rise and 33 to 80 percent loss with a two-meter rise, 90 percent of which would be in the southeastern United States (Titus 1989). In California, 35 to 100 percent of the EPA study wetlands were projected to be lost during the same period. This loss could be reduced to 1 to 18 percent if developed or protected areas were abandoned to allow landward migration of wetlands (Armentano, Park, and Cloonan 1988; Titus 1989).

General possible impacts of global warming on estuaries include:

- the salinity gradient from saline to fresh water will be geographically displaced landward, and currently fresh waters will become more saline;
- some habitat will be lost where landward migration of estuaries is prevented by geographic or artificial barriers;
- conflicts between encroaching estuaries and other competing land-uses will increase, leading to a net loss of estuarine habitat;
- many rare and protected plant populations (even species) will probably be lost;
- the incidence of heat and desiccation on species will increase.

In California, estuaries can be grouped into several major types: 1) structural basins (e.g., Carpinteria Salt Marsh and Goleta Slough) that are often hypersaline, have steep watersheds, and receive much sediment; 2) mouths of canyons that support small lagoons (e.g., Devereaux Slough) with great seasonal fluctuations in salinity and water regimes; 3) mouths of rivers (e.g., the Sacramento/San Joaquin Delta) with brackish lagoons and contiguous riparian corridors; and 4) large bays with extensive salt and brackish marshes (e.g., San Francisco Bay) and contiguous tidal riverine and palustrine habitats. Each of these estuaries may be affected differently by a rise in sea level, depending on natural and artificial barriers that would obstruct the landward migration of wetlands and saline water.

RIVERINE SYSTEMS

The riverine systems include wetlands and deepwater habitats contained within channels, except those supporting woody or persistent emergent vegetation (=palustrine) or those with water containing ocean-derived salts in excess of 0.5 percent (=estuarine) (Cowardin et al. 1979). The distribution of some species (particularly fish) is sensitive to water salinity. Areas such as the Sacramento River are important for spawning of anadromous fish species.

Possible Impacts of Climate Change

The global warming scenario for California could have profound effects, which may, however, be subtle or difficult to interpret. Potential effects include: 1) increased winter runoff and decreased spring and summer runoff from mountains; 2) increase in salinity and inundation from sea-level rise; 3) increase in evaporation as a result of increased temperatures; and 4) movement of saline water landward in riverine systems.

Runoff. Warmer temperatures will melt snow earlier and cause more of the precipitation to occur as rain, which could cause a reduction in snowpack, an increase in winter runoff, and a reduction in spring and summer flows (Gleick 1988; Smith and Tirpak 1989; Smith 1989). Because the current reservoir system in the Central Valley does not have the capacity to store more winter runoff and also provide adequate flood control, the additional winter flow would have to be released, causing increased winter erosion and decreased spring and summer inundation of riverine wetlands.

Increased Water Demand. A temperature increase of 4° C with no increase in rainfall is estimated to produce an average reduction of 10 percent in annual runoff in northern California due to higher evaporation rates (Gleick 1988; California Energy Commission 1989) and as much as a 62 percent decrease in summer runoff. An increase in temperature also will increase demand for water resources for agriculture and urban needs (California Energy Commission 1989), which will further stress riverine systems, particularly if the timing and quantity of runoff change. All of these stresses could affect the species composition, productivity, fishery, and general environmental quality of riverine wetlands and adjacent deepwater habitats in California.

Increased Salinity. Stress in wetland vegetation as a result of increased salinity from a rise in sea level may alter species distributions and successional patterns of plant communities (Pezeshki, DeLaune, and Patrick 1987). Reduced spring and summer flows and a potential decrease in precipitation would increase the potential for movement of salt water into the riverine wetlands.

Competition between human uses of water resources and requirements to maintain natural ecosystems could result in serious conflicts. To preserve the freshwater nature of riverine and adjacent palustrine wetlands, water may have to be released from reservoirs to augment reduced spring and summer flows of rivers that empty into estuaries, thereby forcing salt and brackish water back into bays (Titus 1989). However, potentially increased human demands on water resources, partly from the anticipated higher temperatures, may make this alternative impossible. Construction of new reservoirs to provide additional holding capacity to capture potential increased winter runoff would cause the destruction of other riverine habitats where the reservoirs would be built, and could further reduce sediment input to areas such as the Sacramento/San Joaquin Delta, limiting the potential for estuarine wetlands to keep pace with a rise in sea level.

Landward Movement of Saline Waters. A rise in sea level by one meter will cause significant movement of salt water up rivers and change tidal riverine wetlands and deepwater habitats into estuarine wetlands and deepwater habitats. This process would be harmful to native plants and animals generally restricted to the freshwater habitats and could seriously threaten human uses of fresh water derived from these sources (Titus 1989). Increased salinity could also prevent spawning of anadromous fish species.

LACUSTRINE SYSTEMS

Lacustrine systems include wetlands and deepwater habitats situated in basins or dammed river channels. They lack woody or persistent emergent vegetation, generally exceed eight hectares, often have wave-formed or bedrock shorelines, in their deepest portions may exceed two meters at low water, and never exceed 0.5 percent salinity from ocean-derived salts (Cowardin et al. 1979). A lacustrine system may be bounded by the estuarine, riverine, or palustrine systems and may include a limnetic subsystem (deepwater habitats) and a littoral subsystem (wetland habitats), which extend from the shore to the maximum depth of growth of nonpersistent emergent plants (Cowardin et al. 1979). Wetlands associated with desert lakes and their lacustrine wetlands are important habitats for many endangered species and for migratory birds.

Possible Impacts of Climate Change

The effects of global warming on lacustrine systems could include 1) increase in temperatures; 2) change in water input and outflow amounts and patterns; and 3) changes in air quality that affect water chemistry.

Temperature. Increased temperatures and decreased rainfall could reduce lake levels and change shoreline vegetation. Where lake levels are low, pollution is diluted less, and water quality could decline, affecting entire lacustrine systems. Higher temperatures in lakes could increase the growth of aquatic plants and algae, change circulation patterns, and reduce levels of dissolved oxygen (Smith 1989; California Energy Commission 1989).

Runoff and Evaporation. Changes in the amount of runoff and evaporation (as described for riverine wetlands) could also affect lacustrine wetlands. Desert lakes of California (e.g., Mono and Owens lakes), already stressed by excessive water diversion and high rates of evaporation, could be further stressed by increased rates of evaporation and decreases in precipitation. Wetlands on the margins of these lakes might desiccate earlier each year and could be reduced in size, composition, and zonation. However, potential increases in winter runoff from nearby mountains could provide beneficial inflows for systems such as Mono Lake.

PALUSTRINE SYSTEMS

The palustrine system includes: 1) all wetlands dominated by trees, shrubs, persistent emergent plants, emergent mosses and lichens; 2) all such wetlands as occur in tidal areas where salinity due to ocean-derived salts is less than 0.5 percent; and 3) all nonvegetated wetlands less than eight hectares without wave-formed shorelines and where salinity due to ocean-derived salts is less than 0.5 percent (Cowardin et al. 1979).

In California, palustrine wetlands include, for example, small bodies of open water (e.g., dune swale ponds), nonpersistent emergent wetlands (e.g., vernal pools), persistent emergent wetlands (e.g., freshwater and alkali marshes and seeps), scrub/shrub wetlands (e.g., riparian and alkali scrub), and forested wetlands (e.g., riparian woodlands and forests). These wetlands support many endemic and rare and endangered plants and animals, are important for migratory birds, often have enormous food chain support values, and are highly regarded for their scientific, educational, recreational, and aesthetic values.

Possible Impacts of Climate Change

Palustrine wetlands could be affected by all aspects of global warming, including increases in temperature, changes in precipitation, a rise in sea level, and changes in air quality. Both negative and beneficial impacts might occur. For example, in discussing potential benefits of global warming, the California Energy Commission (1989) has suggested that an increase in rainfall in mountainous and foothill areas could lead to increased soil-moisture levels and an increase in the rangeland yield of meadows, which are generally forms of palustrine emergent wetland. They emphasize, however, that droughtlike conditions may become more frequent in some areas and "desertification" might affect others. Such changes could result in the loss of palustrine wetlands, an increase in alkaline soil conditions, a displacement of wetland communities, and regional extirpation of sensitive species. Specific impacts are best described in reference to specific examples of palustrine wetlands.

Desert Marshes and Alkali Flats—Fish Slough. Fish Slough in Owens Valley supports extensive flooded marshes, perennial springs, sensitive alkali flats, endemic and/or endangered species of fish, plants, and a snail, and is valued highly for its aesthetics and biological and cultural resources (Bureau of Land Management 1985; Forbes, Ferren, and Haller 1988). Changes in precipitation, snowmelt, flow from springs, and evapotranspiration rates due to global warming could seriously alter the sensitive balance of this ecosystem.

Central Valley Freshwater Marshes. Extensive natural and artificial marshes occur in the Central Valley (e.g., Gray Lodge Wildlife Management Area in the Sacramento Valley) and are major habitat areas for migratory birds. Henderson (1989) suggests that increased temperatures and seasonal changes in rainfall of inland areas could reduce such wetlands in North America and seriously affect the reproduction of breeding waterfowl. Gleick (1988) suggests that changes in wildlife refuges of the Central Valley could affect the last remaining stopovers for migratory waterfowl in the Pacific Flyway. Increased runoff from the Sierra Nevada, however, could be managed to offset regional changes in climate and could result in the preservation or increase in such wetlands, unless conflicts develop over the use of water resources for human needs. Such a conflict is likely because 85 percent of the state's water supplies already are used for irrigation (Smith 1989).

Vernal Pools.   Vernal pools, among California's most rare and sensitive wetland habitats, are seasonal wetlands in depressions that support many endemic and endangered organisms with winter and spring rains, followed by summer desiccation. Because of their occurrence on relatively flat terrain, vernal pools in general have received frequent, widespread, and often devastating perturbations due to the spread of agriculture and urban development (Cheatham 1976; Zedler 1987). Increased temperatures and reduced rainfall might result in a loss of shallow pools, decrease in size of deeper pools, loss of species that are growing at the limits of their ranges, and the endangerment of rare, narrowly restricted species.

Coastal Palustrine Marshes.   Various types of palustrine marshes adjacent to or landward of estuaries and/or impounded by levees would be threatened throughout California by a rise in sea level because of increases in inundation and salinity. Particularly vulnerable are those marshes in the Delta and South Bay regions of San Francisco Bay (Williams 1985; Armentano, Park, and Cloonan 1988; California Energy Commission 1989; Josselyn and Callaway 1988). Emergent wetlands dominated by tules and cattails, seasonal alkali wetlands on deltaic deposits, salt ponds, and nontidal brackish marshes in historic estuarine areas could be permanently inundated with a one-meter rise in sea level by the year 2030, especially if the levee system in the Delta region is not maintained.

Riparian Corridors.   California is rich in examples of stream and riverbank vegetation dominated by various trees and shrubs correlated with latitude, elevation, hydrology, or proximity to the coast. Depending on the severity of temperature and precipitation changes or the intensity of air pollution, the species composition of some riparian scrubland, woodlands, or forests could change significantly. Major climatic zones could shift as much as 30–60 km northward (Knox 1989). Because the shift in climate zones might take place in decades rather than thousands of years, many riparian tree species in North America might not be able to migrate northward in latitude or upward in elevation at a rate necessary for germination, establishment, and growth (Roberts 1989; Smith and Tirpak 1989), resulting in some treeless riparian areas.

INTERIOR ECOSYSTEMS

Although much information is available concerning the terrestrial vegetation and climate of California (Barbour and Major 1988), the only comprehensive relationship of vegetation and climate in relation to climate change is presented by Westman and Malanson (1990), from whom our figure 3 is taken. This figure maps the distribution of vegetation types in two-dimensional space defined by mean temperature in the warmest month and annual precipitation. For example, a 4° C warming on a site

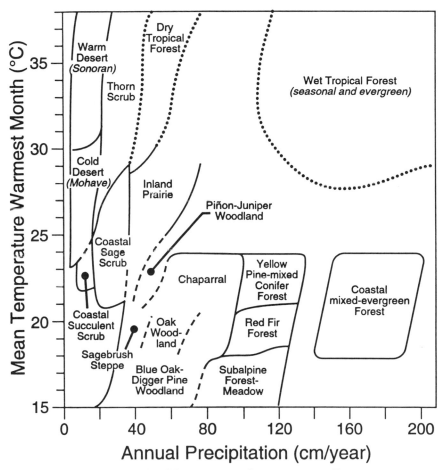

Figure 3. Classification of California vegetation types according to temperature and precipitation. (From Westman and Malanson 1990)

currently anywhere within subalpine forest/meadow space results in mean temperatures that occur nowhere within subalpine forest/meadow currently. This condition suggests that subalpine forest/meadows will be eventually replaced by red fir forest.

In the following sections, six different dominant California interior vegetation types or ecosystems are considered: forests, lakes and streams, shrublands, inland prairies, deserts, and alpine ecosystems of the Sierra Nevada.

FORESTS

What We Know about Forests

Physiology. Physiological processes in trees are greatly influenced by environmental variables. These relationships are well known in general. For example, rates of photosynthesis are related to light levels, moisture stress, temperature, availability of nutrients and CO_2, and other plant stresses (such as air pollution) (Kramer and Kozlowski 1979; Smith 1981). Laboratory studies have provided evidence that drought stress is offset in some species, at least temporarily, by elevated levels of CO_2 (e.g., Kramer and Sionit 1987). However, there is little information available regarding the response of forest-grown trees to the combined effects of elevated temperatures, changed moisture regimes, and enhanced CO_2.

Populations. Population-level studies of forest trees emphasize reproduction and life cycles (e.g., Harper 1977). Trees have different tolerance levels for temperature, light, moisture stress, and nutrients at different stages in their life cycles. For example, seedlings may be girdled and killed by soil-surface temperatures that do not influence mature trees adversely. These tolerance levels may influence tree growth and viability at many stages of the life cycle. Climate change could affect the reproductive cycle, seedling survival, and mortality at many ages from combined stresses and reduction in productivity, chilling requirements for bud break, and hardening off at the end of the growing season (Fowells 1965).

Communities. The relationships between communities and climate are well documented, including those for forest cover types in the United States. Palynological investigations document the relationship between long-term climate change and vegetation change (Faegri and Iversen 1975). The study of past climate and vegetation change in California has lagged behind similar research in other parts of North America. Only a handful of Quaternary scientists have worked in California, and they have been hampered by the large floristic and topographic diversity of California (Adam 1985). Adam has inventoried the sites with fossil pollen records from California. Over one hundred continuous records exist for California, but the striking conclusion from Adam's survey is that many of the best data are not published.

Ecosystems. Natural, undisturbed ecosystems, their processes and dynamic equilibria, vary with vegetation, soil, and climate. Ecosystem processes (e.g., decomposition) occur at different rates in different ecosystems as a result of the varying influences of vegetation (and other organisms), soil, and climate. Stressors that change the rate of ecosystem processes will result in changes in the community and changes in the equilibria of the ecosystem. Forest ecosystem processes are generally well understood owing to the intensive studies conducted at several experimental forests in the U.S. Notably, no similar long-term research sites now exist in California. Furthermore, there have been no experiments conducted anywhere which address forest productivity (as distinguished from seedling productivity) or forest ecosystem carbon balance (as distinguished from seedling or very small tree carbon balance). Modeling efforts (e.g., Botkin and Nisbet 1990) currently addressing this issue are hampered by a lack of information about the interrelationships between climate change, CO_2 fertilization, resource limitation, stresses, and tree physiology.

Possible Impacts of Climate Change

Specific impacts of global warming on California forests include:
- elimination of Douglas fir from lowlands due to loss of winter chill conditions needed for germination and growth (Leverenz and Lev 1987);
- increased fuel loading due to Douglas fir mortality;
- widespread movement of forests upslope from their present locations (California Energy Commission 1989);
- increased productivity of redwood in the northern areas, but reduction of redwood range in southern areas if precipitation decreases;
- increased rates of decomposition and nutrient release from litter;
- succession in forest gaps by shrub species such as ceanothus and alder, leading to an increase in spatial vegetation diversity;
- wildlife habitat changes that will favor species which take advantage of gaps and dead wood;
- likely deterioration of stream-water quality due to increased rates of decomposition, weathering, and erosion caused by tree mortality.

LAKES AND STREAMS

What We Know about Lakes and Streams

Aquatic ecosystems integrate coupled atmospheric and terrestrial watershed effects in their physical, chemical, and biological characteristics. Aquatic ecosystems, particularly in mountain environments, are quite sensitive to environmental disruptions. At high elevations (alpine and subalpine) snowmelt quantity and chemistry have immediate effects on stream and lake dynamics (Melack et al. 1989). Many lakes are flushed annually during spring snowmelt. They are naturally very oligotrophic and of low alkalinity. As a consequence, they respond quickly to snowmelt acidity (Melack et al. 1989) and nutrient dilution or enrichment from the watershed or directly from the atmosphere (Goldman, Jassby, and Powell 1989). Interannual variation in the heat budgets (Strub, Powell, and Goldman 1985) and primary productivity (Goldman, Jassby, and Powell 1989) of subalpine lakes are strongly linked to climatic variation. Mountain streams are often almost completely dependent on the surrounding forest for their source of reduced carbon, as aquatic productivity is often light- and nutrient-limited (Meyer, Likens, and Sloane 1981). Any disturbance of a watershed can have both immediate and long-term effects on the nutrient and sediment load of the stream (Vitousek and Melillo 1979).

Possible Impacts of Climate Change
Examples of likely impacts include:
* changes in the annual balance and seasonality of snow and rain, which may greatly affect the pattern and quantity of nutrient and sediment release from watersheds (Leonard et al. 1979; Lewis and Grant 1980);
* disruption of seasonal flow regimes, which may severely reduce the survival of fish and amphibians that require in-stream incubation of their eggs (Cooper, Jenkins, and Soiseth 1988);
* likely decreases of water supply by warming (see Gleick 1988), leading to depauperization of aquatic populations;
* change in seasonal runoff pattern, which could decrease fish-egg survival both through increased winter/spring sedimentation and early fall drying of the spawning areas;
* exacerbated eutrophication and fish death in high elevation freshwater ecosystems.

SHRUBLANDS

What We Know about Shrublands
The only information available on the direct effect of elevated carbon dioxide on chaparral species derives from measurements of photosynthetic response to elevated carbon dioxide and temperature in leaf chambers placed on field plants (Wright 1974). This study indicates that some chaparral species are able to use additional CO_2 to stimulate photosynthetic rates. The long-term effects of elevated CO_2 on growth of the species alone, or in combination with climatic changes, are not

known. There have been no long-term chamber or field-exposure studies done with elevated CO_2 on any of the California shrubland species.

Possible Impacts of Climate Change

Specific impacts on shrublands may include:

- invasion of chaparral areas by piñon-juniper woodland and dry tropical forest under global warming conditions without changes in precipitation (suggested by figure 3);
- elevated ozone levels and heightened fire intensity induced by increasing fuel loads under altered climate, which may lead to extirpation of the dominant species within a century (Westman and Malanson 1990);
- coastal sage scrub encroachment into areas away from the immediate coast, particularly in the Los Angeles Basin (Westman and Malanson 1990);
- enhanced chaparral growth under increased precipitation, or, alternatively, greater drought stress under lower precipitation with increased mortality leading to increased fire frequency.

INLAND PRAIRIES

What We Know about Inland Prairies

Before its agricultural development, the Central Valley of California was occupied by a mixture of grassland, dry shrubland, and riparian woodland. Increased fire frequency induced by Native Americans, combined with domesticated grazing introduced by settlers in the nineteenth century, encouraged a relative abundance of grasses, especially annual grasses, in this vegetative mix. The dry shrubland was probably characterized by such semiarid species as *Artemisia tridentata* and *Haplopappus linearifolius*. Riparian woodland would have been dominated by such species as cottonwood, *Populus trichocarpa*.

Possible Impacts of Climate Change

Specific likely impacts on shrubland ecosystems include:

- expansion of the dry shrubland and grassland elements into the chaparral and oak and digger pine woodlands of the interior foothills (Westman and Malanson 1990);
- probable migration of valley riparian vegetation along foothill watercourses as net evapotranspiration stress increases.

DESERTS

What We Know about Deserts

One-third of California is desert, including all of the area inland of the Transverse and Peninsular ranges in southern California, the valleys east of the Sierra and Cascade ranges, and part of the southern San Joaquin Valley. Except for the San Joaquin Valley, the deserts of California are extensions of larger desert regions that cover much of the southwestern United States and northwestern Mexico. These deserts have different climates and support different plant and animal communities. The Sonoran Desert extends into southeastern California at elevations below 500 m. The California component of the Sonoran Desert is extremely arid, with a total annual rainfall of only 50 to 150 mm. Consequently, most of the region is characterized by a sparse vegetation of creosote bush, bursage, and occasionally abundant ephemerals (Burk 1977). Only in washes and on rocky slopes of desert mountains are found large cacti and arborescent species characteristic of the wetter parts of the Sonoran Desert in Arizona and Sonora, Mexico. To the north of the Sonoran Desert elevation increases, and above about 500 m (approximately along the Riverside–San Bernardino county line) there is a gradation into the Mojave Desert. The principal climatic features of the Mojave Desert are an extreme deficit of summer rainfall, more winter rainfall than the Sonoran Desert, and nightly frost in midwinter (January mean monthly minimum temperature below 0° C). Creosote bush and bursage remain the dominant plant species in the Mojave (Vasek and Barbour 1977), but the associated arborescent species differ from those of the Sonoran Desert. Many of the Sonoran Desert species are frost-sensitive and apparently cannot survive in the Mojave Desert (Turner and Brown 1982).

To the north of the Mojave Desert, lying in the vast rain shadow of the Sierra and Cascade mountain ranges, is the extreme western edge of the Great Basin Desert. Its high elevation with mean minimum temperatures below 0° C during the winter make this "cold" desert botanically distinct from the "hot" deserts (MacMahon 1979). Big sagebrush and shadscale are principal dominant shrubs with bunchgrasses in more moist habitats (Young et al. 1977). Moisture comes mostly during the coolest half of the year; summers are hot and dry. Thus, periods of suitable moisture for plant growth are substantially out of phase with periods of suitable temperature, greatly limiting plant growth.

Survival in the desert depends on an organism's ability (adaptation) to utilize specific "pulses" or "windows" of resource availability (Noy-Meir 1973). Winter annual plants, for example, are adapted to germinating, growing, and completing their life cycle within the winter season, while summer annual species are adapted to the short summer season. Desert granivores are adapted to utilizing seeds of a select group of species. To the extent that climate change brings about a restriction of availability of a resource, say summer moisture or seeds, then species adapted to utilizing that resource will incur direct impacts, with potential decline in vigor and abundance, or even elimination.

Possible Impacts of Climate Change

Specific likely impacts in California deserts include:

- acceleration of growth during winter and spring when temperature is the limiting growth factor;
- increased evapotranspiration rates, leading to mortality of more mesic species during dry periods and replacement by more drought-tolerant species;
- interaction of other factors (e.g., grazing, tillage, erosion) causing other directional shifts in vegetation distribution;
- general decline in vegetation cover on areas of low topographic relief, leading to increased dune formation in some areas, increased wind erosion, and soil transport;
- possible local extinction of some plant populations as warming and drying conditions push their climatic ranges beyond the tops of existing mountains;
- increased species diversity in the Sonoran and parts of the Mojave deserts due to increased precipitation, creating a more certain summer growth "window";
- shifts in species composition from water-use-efficient forms to nutrient-use-efficient forms, if desert plant productivity is more nutrient-limited than moisture-limited (Schulze and Chapin 1987);
- increase in winter minimum temperatures allowing some incursion of frost-sensitive Sonoran Desert species (e.g., saguaro cactus) northward into the Mojave, particularly if summer moisture increases.

ALPINE ECOSYSTEMS OF THE SIERRA NEVADA

What We Know about Sierra Alpine Ecosystems

Most of California's alpine ecosystems are found in the Sierra Nevada above about 2,800 meters. These systems may be generally defined as small lake and stream watersheds having dilute surface waters, acidic soils, exposed bedrock, and sparse vegetation. The hydrology of these watersheds is dominated by snowfall, which can account for more than 90 percent of the wet deposition falling above the forest line. Most of our understanding of ecological processes in these watersheds and surface waters is based on studies to investigate the effects of acid deposition and other anthropogenic disturbance (e.g., Cooper, Jenkins, and Soiseth 1988).

Possible Impacts of Climate Change

The following changes in alpine ecosystems may be expected:

- increased incidence of drought as summer temperatures increase, with increased rates of evapotranspiration, leading to decreased plant production and decreased soil microbial activity;

- a longer period of open water in lakes and streams, leading to warmer surface waters, longer stratification periods, greater algal blooms (depending on nutrient limitation), and decreases in fish production;
- early drying of seasonal ponds affecting amphibian production;
- increase in ozone concentrations during the growing season resulting in injury to plants;
- increase in the incidence of extreme events such as flooding in winter and more avalanche scouring due to wet snow events.

RECOMMENDATIONS FOR RESEARCH

It is obvious that global climate change of the magnitude suggested in foregoing scenarios will result in very serious impacts on California natural ecosystems. It is also obvious that we lack sufficient information both to estimate the impacts with certainty and to know how to respond effectively. California needs an aggressive ecosystem research program to 1) compile existing sources of information; 2) determine accurate models for predicting regional changes; 3) begin monitoring of current conditions to provide a baseline for future comparisons; and 4) conduct experiments under field conditions and in the laboratory to better understand biological feedback mechanisms.

The best approach to understanding potential problems and setting policy regarding the impacts of global warming on the California environment is to develop an integrated statewide program of ecological study. We suggest that the University of California in conjunction with state agencies develop a *California Ecosystems Program* to foster collaboration on natural systems research. The state at present has only a few small long-term research sites in terrestrial ecosystems (e.g., Deep Canyon, administered by UC Riverside), and it has no large laboratory research facility for manipulation of plant communities at the level required. A *California Ecosystems Program* could serve as a clearinghouse and storage bank for data, coordinate specific research projects and activities on key habitats, and provide integrative communication mechanisms such as sponsorship of symposia and regular publications.

The mission of a *California Ecosystems Program* should include these general areas of activity:

1) **Collect, collate, and analyze existing data/databases and integrate them with remote imagery and geographic information systems (where appropriate to produce cartographic products) to use as a guide in planning additional research and to produce a baseline for impact analysis.** Much information on California ecosystems exists in unanalyzed data sets (i.e., forty years of California Coastal Fisheries monitoring data; ocean cores with fish scale data; over one hundred pollen data

sets/diagrams; USC data on ocean anoxia). Much floristic, faunistic information exists in herbaria, museums, and botanic gardens. Other biological survey information exists in environmental impact reports and reports on coastal lands. Several large-scale statewide programs, such as the Natural Diversity Database, can provide valuable information to help fill baseline database requirements for impact analysis. New research programs should be designed to provide the information necessary to support present and planned predictive models.

2) **Design the general structure of ecosystem models necessary to predict ecosystem response to climate change.** The complex, multifactor relationships between ecosystems and climate require the use of models as integrative repositories of species/climate information. For proper analysis of effects of climate change, many factors must be allowed to co-vary in order to predict future ecosystem states. Predictive ecosystem models may require relatively intensive data on species productivity, community structure and diversity, ecotypic variation, nutrient cycling, and energy flow patterns.

3) **Design and implement large-scale monitoring programs for major marine, coastal, and inland ecosystems to gather data necessary to support simulation models.** For example, the California Cooperative Oceanic Fisheries Investigation (CalCOFI) has monitored plankton species presence/abundance and ocean physical parameters quarterly for the last forty years in the waters of the California Current. Other large-scale, long-term monitoring programs should be instigated that provide high-frequency sampling of:

- physical factors (temperature, humidity, light intensity, wind speed/strength/direction, etc.);
- chemical and nutrient factors (salinity, DO, BOD, pH, N, P, K, etc.);
- biological factors (chl$_a$, species presence/abundance, etc.);
- geological/paleontological factors (soil moisture, fertility, pollen deposition, etc.).

Data from these monitoring programs should be related to large-scale ocean and terrestrial processes to elucidate mechanisms that can be incorporated into predictive models.

4) **Design process studies that will support the construction of predictive models.** Small-scale and large-scale field and laboratory experimental programs should be designed to explain responses to climate change of ecosystem processes in a manner suitable for incorporation into predictive models. One important area of investigation would be to determine relationships in time and space between scales of the climate and vegetation models and scales of ecosystem response. Also, ENSO events and patterns seen in earlier interglacials could be used as possible analogs of what might happen in the next century.

5) **Plan for the preservation, restoration, and creation of wetlands.**

Wetlands are particularly important natural areas in California for supporting migratory fowl, unique or endangered species, and recreation. Ecosystem research should be geared toward preservation efforts to save wetlands most sensitive or ecologically significant. Research should also evaluate the reconstruction of wetlands in new areas as sea level rises.

REFERENCES

Adam, D. P. 1985. Quaternary pollen records from California. In *Pollen records of late-Quaternary North American sediments*, ed. V. M. Bryant, Jr., and R. G. Holloway, 125–140. Dallas: American Association of Stratigraphic Palynologists Foundation.

Armentano, T. V., R. A. Park, and C. L. Cloonan. 1988. Impacts on coastal wetlands throughout the United States. In *Greenhouse effect, sea level rise, and coastal wetlands*, ed. J. G. Titus. EPA 230-05-86-013. Washington, D.C.: U.S. Environmental Protection Agency.

Barbour, M. G., and J. Major, eds. 1988. *Terrestrial vegetation of California*. New expanded ed. California Native Plant Society, Special Publication no. 9. Davis, Calif.: California Native Plant Society.

Botkin, D. B., and R. A. Nisbet. 1990. Projecting the effects of climate change on biological diversity in forests. In *Consequences of the greenhouse effect for biodiversity*, ed. R. Peters. New Haven: Yale University Press.

Bureau of Land Management. 1985. *Management plan for Fish Slough: Area of critical environmental concern, a cooperative management program*. Department of the Interior, Bishop Resource Area.

Burk, J. 1977. Sonoran Desert. In *Terrestrial vegetation of California*, ed. M. G. Barbour and J. Major, 869–889. New York: Wiley.

Byron, E. R., A. Jassby, and C. R. Goldman. 1988. Water quality of subalpine lakes. In *The potential effects of global climate change on the United States*, ed. J. B. Smith and D. A. Tirpak. EPA-230-05-050. Washington, D.C.: U.S. Environmental Protection Agency.

California Energy Commission. 1989. *The impacts of global warming on California: Interim report*. P500-89-044. Sacramento: California Energy Commission.

Cheatham, N. D. 1976. Conservation of vernal pools. In *Vernal pools: Their ecology and conservation*, ed. S. Jain. Institute of Ecology Publication no. 9. Davis, Calif.: Institute of Ecology, University of California, Davis.

Chelton, D. B. 1981. Interannual variability of the California Current: Physical factors. *CalCOFI Report* 22:34–48.

Chelton, D. B., P. A. Bernal, and J. A. McGowan. 1982. Large-scale physical and biological interaction in the California Current. *J. Mar. Res.* 40:1095–1125.

Cooper, S. D., T. M. Jenkins, and C. Soiseth. 1988. *Integrated watershed study: An investigation of fish and amphibian populations in the vicinity of the Emerald Lake Basin, Sequoia National Park: Final report*. Sacramento: California Air Resources Board.

Cowardin, L. M., V. Carter, F. C. Golet, and E. T. LaRoe. 1979. *Classification of wetlands and deepwater habitats of the United States.* FWS/OBS-79/31. Washington, D.C.: Office of Biological Services, U.S. Fish and Wildlife Service.

Davis, M. B. 1989. Lags in vegetation response to global climate change. *Climate Change* 15:75–82.

Faegri, K., and J. Iversen. 1975. *Textbook of pollen analysis.* New York: Hafner Press.

Forbes, H. C., W. R. Ferren, Jr., and J. R. Haller. 1988. The vegetation and flora of Fish Slough and vicinity, Inyo and Mono counties, California. In *Plant biology of eastern California,* ed. C. A. Hall and V. Doyle-Jones. Natural History of the White-Inyo Range Symposium, vol. 2. Los Angeles: University of California White Mountain Research Station.

Fowells, H. A. 1965. *Silvics of forest trees of the United States.* Agriculture Handbook no. 271. Washington, D.C.: U.S. Department of Agriculture, Forest Service.

Gleick, P. H. 1988. Impacts on natural resources: Climate change and its impact on water resources. Testimony prepared for Subcommittee on Water and Power Resources, Fort Mason Conference Center, October 17, 1988.

Goldman, C. R., A. Jassby, and T. Powell. 1989. Interannual fluctuations in primary production: Meteorological forcing to two subalpine lakes. *Limnol. Oceanogr.* 34:310–323.

Harper, J. L. 1977. *Population biology of plants.* New York: Academic Press.

Henderson, S. 1989. How it could be: Species. *EPA Journal* 15, no. 1:21–22.

Hickey, B. M. 1979. The California Current system—hypotheses and facts. *Prog. Oceanog.* 8:191–279.

Josselyn, M., and J. Callaway. 1988. *Ecological effects of global climate change: Wetland resources of San Francisco Bay.* Prepared for the Environmental Protection Agency, Environmental Research Laboratory, Corvallis, Oregon.

Knox, J. B. 1989. Climate scenarios selected for the first UC/DOE Workshop. Backgrounder Contribution no. 18.

Kramer, P. J., and T. T. Kozlowski. 1979. *Physiology of woody plants.* New York: Academic Press.

Kramer, P. J., and N. Sionit. 1987. Effects of increasing carbon dioxide concentration on the physiology and growth of forest trees. In *The greenhouse effect, climate change, and U.S. forests,* ed. W. E. Shands and J. S. Hoffman, 219–246. Washington, D.C.: The Conservation Foundation.

Leonard, R. L., L. A. Kaplan, J. F. Elder, R. N. Coats, and C. R. Goldman. 1979. Nutrient transport in surface runoff from a subalpine watershed, Lake Tahoe basin. *California Ecol. Monogr.* 49:281–310.

Leverenz, J. W., and D. J. Lev. 1987. Effects of carbon dioxide-induced climate changes on the natural ranges of six major commercial tree species in the western United States. In *The greenhouse effect, climate change, and U.S. forests,* ed. W. E. Shands and J. S. Hoffman, 123–188. Washington, D.C.: The Conservation Foundation.

Lewis, W. M., and M. C. Grant. 1980. Relationships between snow cover and winter losses of dissolved substances from a mountain watershed. *Arctic and Alpine Research* 12:11–17.

MacMahon, J. A. 1979. North American deserts: Their floral and faunal components. In *Aridland ecosystems: Structure, functioning and management,* ed. D. W. Goodall and R. A. Perry, 1:21–82. London: Cambridge University Press.

McGowan, J. A. 1985. El Niño in 1983 in the southern California Bight. In *El Niño north*, ed. W. W. Wooster and D. L. Fluharty. Seattle: Washington Sea Grant Program, University of Washington.

Melack, J. M., S. D. Cooper, T. Jenkins, L. Barmuta, S. Hamilton, K. Kratz, J. Sickman, and C. Soiseth. 1989. *Chemical and biological characteristics of Emerald Lake and streams in its watershed, and the responses of the lake and streams to acidic deposition: Final report*. Prepared for the California Air Resources Board. Santa Barbara, Calif.: Marine Science Institute and Dept. of Biological Sciences, University of California.

Meyer, J. L., G. E. Likens, and J. Sloane. 1981. Phosphorus, nitrogen, and organic carbon flux in headwater streams. *Arch. Hydrobiol.* 91:28–44.

Noy-Meir, I. 1973. Desert ecosystems: Environment and producers. *Ann. Rev. Ecol. Sys.* 4:25–51.

Pezeshki, S. R., R. D. DeLaune, and W. H. Patrick, Jr. 1987. Effect of salt water intrusion on physiological process in wetland plant communities. In *Wetland and riparian ecosystems of the American West*, ed. K. M. Mutz and L. C. Lee. Denver: Planning Information Corp.

Roberts, L. 1989. Research news: How fast can trees migrate? *Science* 243: 735–737.

Roos, M. 1989. Possible climate change and its impact on water supply in California. In *Oceans '89: An international conference . . . , September 18–21, 1989, Seattle, Washington, USA*. Piscataway, N.J.: IEEE Service Center.

SFBCDC (San Francisco Bay Conservation and Development Commission). 1987. *Future sea level rise: Predictions and implications for San Francisco Bay*. Prepared by Moffatt and Nichol, Engineers; Wetland Research Associates, Inc. San Francisco: San Francisco Bay Conservation and Development Commission.

Schulze, E. D., and F. S. Chapin, III. 1987. Plant specialization to environments of different resource availability. In *Potentials and limitations of ecosystem analysis*, ed. E. D. Schulze and H. Zwolfer, 120–148. Ecological Studies, 61. New York: Springer-Verlag.

Simpson, J. J. 1984. El Niño–induced onshore transport in the California Current during 1982–83. *Geophys. Res. Let.* 11:232–236.

Smith, J. B., and D. A. Tirpak, eds. 1989. Potential impacts of climate warming on California. In *The potential effects of global climate change on the United States*, vol. 1, *Regional studies*, chap. 4. EPA-230-05-89-050. Washington, D.C.: U.S. Environmental Protection Agency.

Smith, J. E. 1989. How it might be: Water resources. *EPA Journal* 15, no. 1:19–20.

Smith, W. H. 1981. *Air pollution and forests: Interactions between air contaminants and forest ecosystems*. New York: Springer-Verlag.

Strub, P. T., T. Powell, and C. R. Goldman. 1985. Climatic forcing: Effects of El Niño on a small, temperate lake. *Science* 227:55–57.

Titus, J. G. 1989. How it could be: Sea levels. *EPA Journal* 15, no. 1:14–16.

Turner, R. M., and D. E. Brown. 1982. Sonoran desert scrub. *Desert Plants* 4:181–221.

Vasek, F. C., and M. G. Barbour. 1977. Mojave desert scrub vegetation. In *Ter-*

restrial vegetation of California, ed. M. G. Barbour and J. Major, 835–867. New York: Wiley.

Vitousek, P. M., and J. M. Melillo. 1979. Nitrate losses from disturbed forests: Patterns and mechanism. *Forest Sci.* 25:605–619.

Westman, W. E., and G. P. Malanson. 1990. Effects of climate change of Mediterranean-type ecosystems in California and Baja California. In *Consequences of the greenhouse effect on biodiversity*, ed. R. Peters. New Haven: Yale University Press.

Williams, P. B. 1985. *An overview of the impact of accelerated sea level rise on San Francisco Bay*. Philip Williams and Associates Project no. 256. San Francisco: Philip Williams and Associates.

Wright, R. D. 1974. Rising atmospheric CO_2 and photosynthesis of San Bernardino mountain plants. *Amer. Midl. Nat.* 91:360–370.

Young, J. A., R. A. Evans, and J. Major. 1977. Sagebrush steppe. In *Terrestrial vegetation of California*, ed. M. G. Barbour and J. Major, 763–796. New York: Wiley.

Zedler, J. B. 1982. *The ecology of southern California coastal salt marshes: A community profile*. FWS/OBS-81/54. Washington, D.C.: U.S. Fish and Wildlife Service, Biological Services Program.

Zedler, P. H. 1987. *The ecology of southern California vernal pools: A community profile*. Biological Report 85. Washington, D.C.: U.S. Fish and Wildlife Service.

EIGHT

Global Warming from an Energy Perspective

Allen G. Edwards

Global climate change and energy are integrally related. The majority of greenhouse gas emissions are the result of energy production and use; at the same time, warming will affect energy patterns in California through physical increases in energy demand, physical changes in energy supply, and changes in both energy end-use patterns and supplies resulting from climate-change policies.

In 1988 the California Energy Commission, with the help of six other state agencies and the University, began a serious inquiry into the impacts of climate change on California. This inquiry included an analysis of both energy demand and supply impacts. The Commission used its end-use demand forecasting model to project how energy demand would change by the year 2050 from a three-degree Celsius warming. Along with the State Department of Water Resources, the Commission analyzed changes in energy supply that could result from the same-magnitude warming. The findings of the study, which are presented in the sections below, show that climate warming could increase energy demand while reducing energy supply.

There seems to be a growing political consensus that the world (as well as the state) needs to act soon to minimize further commitment to future warming. While California is not likely to experience the physical changes resulting from a warmer climate for years or perhaps decades, policy responses to the warming issue may cause more immediate impacts. This chapter will discuss how policy response to potential warming may be the most significant early impact of the issue on California's energy system. Makers of energy policy face the dilemma of deciding how to respond to the climate warming issue in the face of scientific uncertainties about its timing and seriousness. The chapter will conclude by presenting a conceptual framework for dealing with this dilemma, along with general recommendations for action.

ENERGY DEMAND

Higher temperatures are expected to increase energy demand. An effective doubling of CO_2 (expected to occur sometime between the years 2030 and 2070) would increase the annual average temperature of the state by two to four degrees Celsius (C). As a result, there may be some decreases in winter heating requirements; but because much of California's population resides in warm regions, increases in summer air-conditioning loads would far exceed any decreases in heating load. According to the California Energy Commission, changes in heating and air conditioning from a three-degree C increase in average temperature in the year 2050 would increase peak demand by 3 to 7 percent and overall energy demand by 1.5 to 2.5 percent, assuming no significant change in population (Edwards et al. 1989).

Air-conditioning loads would markedly increase in agriculture and industry as well as in residential and commercial buildings. There would, for example, be an increased need for cooling poultry operations during the warm season, and more energy would be required to chill and freeze food products. Industrial, commercial, and residential buildings would require cooling to make them workable and livable.

Water pumping would account for another major source of increased energy demand. As California's climate warms, the state's supply of surface water is expected to decrease owing to both decreases in snow-pack storage and increased evaporation. At the same time, higher temperatures during growing seasons will increase irrigation water needs for most crops. With traditional water supplies declining and water needs rising, the state will have to rely on additional energy-consuming groundwater pumping to help fill shortfalls. In addition, some of the water conservation methods available to farmers require replacing current open irrigation systems with enclosed systems, which also require more energy for pumping. Agricultural energy demand, of which water pumping is the major component, is currently responsible for about 8 percent of energy demand in the PG&E utility service area (Jaski et al. 1989). Statewide, the Energy Commission estimates that, with a one-degree warming by the year 2010, electricity demand for water pumping could increase by about 1 percent (Baxter and Calandri 1989), and it is expected that this demand will continue to increase well into next century as the climate warms. Many agricultural operations may eventually face the decision of whether to remain in an area where water and energy costs are increasing, or move their operations (where possible) and contend with higher transportation costs.

CHANGES IN ENERGY SUPPLY

Several physical changes in energy supply would result from a global climate warming: changes in hydroelectric generation, in the efficiency of conventional fossil-fueled power plants, and in renewable energy resources.

Hydroelectric Generation

Changes in watershed runoff would directly affect hydroelectric generation. While there is no consensus on whether climate warming will increase or decrease annual precipitation, scientists generally agree that it will decrease snowpack storage in the state's mountain watersheds. As a result, more water will flow in rivers during winter and less during the summer; overall, however, less water will be available for power generation. For example, estimates suggest that in the Feather River Basin a three-degree warming would reduce hydroelectric capacity and energy production by up to 7 percent (Edwards et al. 1989). In addition, less hydroelectric power would be available when it is most needed, during the state's summertime peak demand periods. Increasing competition for water brought on by lower supplies and increased agricultural and urban demand, and needs to protect natural habitat, may further reduce hydroelectric output. California already experiences periods of significant competition for available water, particularly during low rainfall years. Because the state has many other sources of power, hydrogeneration is sometimes reduced in order to release more water for downstream uses. This trend may intensify as the climate warms.

Power Plant Efficiency

Warmer ambient temperatures, as would be expected from a climate warming, may reduce the efficiency of power plant cooling systems, resulting in lower net electricity output from many thermal-electric generating plants. Most fossil-fueled, geothermal, and nuclear power plants use water cooling systems to lower the pressure of steam turbine exhausts; since the electricity output of these plants is partly a function of the difference between their turbines' intake and exhaust pressures, a decrease in exhaust cooling efficiency will reduce their output. For example, the North American Electric Reliability Council estimated that during the hot summer of 1988 the United States lost seventeen thousand megawatts thermal capacity as a result of the high temperatures and constraints on cooling water (*Energy Economist* 1988).

Renewable Energy Resources

The picture on renewable energy resources is mixed. Whereas climate warming would reduce output from conventional generating plants, it may increase output from some renewable sources. Some data have suggested that warming-induced changes in California offshore ocean currents will intensify the summertime wind regimes that drive most of California's wind turbines. This could increase power output at existing sites and allow currently submarginal sites to be commercially developed.

Warming might also increase the near-term availability of biomass fuel. Carbon dioxide fertilization could well increase the volume of agricultural residues available for fuel. Changes in the ranges of forest species might also result in the die-off of large numbers of trees, parts or all of which could be salvaged for fuel. However, these same changes in forest composition would lead in the long run to a decline in forest biomass. For example, the EPA estimates that the western slope of the Sierras could ultimately have only 60 percent of current biomass as a result of a CO_2 doubling (Smith and Tirpak 1988).

ENERGY POLICY IMPACTS

As described above, the physical impacts of global warming will have a significant, though not devastating, effect on California's energy demand and supply. The impacts in other areas—agriculture, water resources, natural ecosystems, forestry, human health, and the overall economy—may be much more intense than those on energy systems. All these impacts should be intermittent and mild in the near term but would intensify over the next several decades. Californians need to recognize that policies to minimize long-term warming, whether they originate from federal or state government, may have a much more immediate impact on the state than the warming itself. The following discussion identifies some of the most likely policy actions and their near-term implications.

Emissions Fee

One policy proposal with potentially widespread impacts is a fee on fossil carbon emissions (a carbon "tax"). Currently discussed primarily in the context of a national or international policy, this fee would have two purposes: First, to account for global warming-related externalities associated with the use of fossil fuels. In this context, the fee would provide the most benefits if the revenues collected were used to prevent and mitigate the warming. Second, the fee would be an incentive to reduce consumption of fossil fuels.

If applied to the major fossil fuels (coal, oil, and natural gas), the fee would affect virtually every aspect of the state's economy. Gasoline and diesel fuel prices would rise. Electricity generated from natural gas (43 percent of California in-state generation) and coal (most of the electricity imported into the state) (California Energy Commission 1989b) would become more expensive. Fertilizer produced by using natural gas would cost more. Even the price of steel used in imported products like automobiles and refrigerators would go up. Eventually, the cost of most goods would rise, and purchasing habits of businesses and consumers could radically change. Such a carbon tax would encourage consumers to make fuel-use decisions with more complete knowledge of costs, including currently hidden environmental costs.

Most analysts assume that such a tax would reduce overall fuel consumption and, in turn, CO_2 emissions. (Estimates on the amount of reduction for any particular tax level vary widely.) Those reductions would likely come as consumers both increase the efficiency of their energy use (by applying technological improvements) and curtail their fuel-using activities. Thus, while a carbon tax could (depending on how high it is set) lead to increased social efficiency, it would likely have a noticeable affect on the lives of Californians. But even though a tax might raise direct energy costs in the near term, it could increase overall energy efficiency to the point where the cost of services provided by using fossil fuels (moving people from one place to another, raising food, heating and lighting buildings, for example) would go down.

The most significant proposals for a carbon tax are currently being discussed at the national and international level rather than as a state tax. While a widely mandated tax would be more effective in reducing overall carbon emissions, even a unilaterally applied state tax could be beneficial if it were no higher than carbon combustion externalities and the tax revenues were used to reduce CO_2 emissions or to mitigate the damage they cause.

Energy Efficiency Improvement Programs

Improving energy efficiency is probably the most promising approach to reducing fossil carbon emissions. There are ample data showing that government policies reduced, or at least stabilized, fossil-fuel consumption from the mid-1970s to mid-1980s. In California, for instance, state-mandated appliance and building efficiency standards have saved over ten thousand gigawatt hours of electricity and a billion therms of natural gas since their introduction in 1977 (California Energy Commission 1988). Federal corporate average fleet economy standards (CAFE) have increased the average fuel efficiency of new cars sold in the United States from fourteen miles per gallon in 1974 to the current twenty-eight miles per gallon (California Energy Commission 1989a).

The consensus among both environmental and energy analysts is that more aggressive energy efficiency policies could dramatically reduce energy consumption and, in turn, fossil-fuel carbon emissions. While estimates vary, cost-effective efficiency improvements might, at minimum, stabilize fossil-fuel emissions at current levels and may, if more optimistic estimates are true, allow cutting current emissions by up to 75 percent. Cost-efficient measures would result in a net benefit to society— reducing costs while providing the same energy service; this does not mean, however, that particular companies and individuals might not suffer. There is some indication, for example, that the current modest oversupply of generating resources is raising California's per-unit electricity rates (customers are paying for underutilized capital equipment). Increased energy efficiency might exacerbate this problem for a short period. However, most electricity users who aggressively apply energy efficiency measures will see their utility bills decrease, even if the rate per KWH is higher. In the long run, by avoiding construction of new, costly power plants the total costs of the system should decline.

Renewables

Another promising option for reducing carbon emissions is substitution of renewable energy resources for fossil-fueled electric generating plants. California currently has over ten thousand megawatts of renewable generation in operation (including hydroelectric, geothermal, biomass, wind, and solar technologies) producing electricity with little or no net carbon emissions. Many of the renewable technologies have the technical maturity efficiently to displace additional fossil-fueled generation, but most currently produce electricity at a higher direct cost. However, when all the external costs of fossil-fueled plants—indirect costs to society such as government subsidies, energy security, and pollution— are accounted for, renewable resources may turn out to be more cost-effective than fossil-fueled generation. Thus, while a shift to renewable technologies may raise the direct cost of electricity, it may lower the overall societal costs of electricity generation.

Carbon Sequestering

According to some scientists, it may be possible to reduce atmospheric concentrations of CO_2 by sequestering carbon. The primary methods being considered rely on trees to take up and store carbon: reforestation and afforestation, and more intense management of existing commercial forests. There is also some discussion of removing CO_2 from power-plant emissions and storing it in deep ocean areas, but high costs and technical difficulties currently preclude this approach from serious consideration.

In California, extensive areas of currently marginal, rural land could be planted with fast-growing tree species. Large, irrigated tree plantations may not be practical since they could exacerbate competition for scarce water. But there is a significant opportunity for planting drought-tolerant trees in currently unused or underused land. Planting trees in many urban areas would have the dual benefit of sequestering carbon and providing cooling shade to reduce air-conditioning energy consumption. In addition, much of the state's commercial forest land, particularly many of the smaller parcels in absentee ownership, could be

TABLE 1. Multiple Benefits of Carbon Reduction Policies

Policy Measures	Reduce Energy Costs	Reduce Regional Air Pollution	Improve Energy Security	Increase Economic Resources Base	Decrease Greenhouse Gas Concentrations
Improving energy efficiency (in stationary and transportation uses)	Major benefit	Major benefit	Major benefit		Major benefit
Increasing renewable energy resources		Major benefit	Major benefit	Modest benefit	Major benefit
Carbon emission fees		Major benefit	Major benefit		Major benefit
Carbon sequestering			Possible benefit[1]	Modest benefit	Major benefit

[1]Tree plantations could become an ongoing source of biomass fuel while, as a whole, remaining a sump for carbon.

more intensively managed to increase substantially the land's productivity and its standing volume of carbon material. Careful consideration must precede decisions about the kinds of trees that are planted, however. Certain varieties of trees naturally emit substantial amounts of reactive hydrocarbons. For example, approximately one-half of the reactive hydrocarbons in the air over Atlanta, Georgia, come from tree emissions. Since these hydrocarbons can be precursors to photochemical smog, planting trees with high hydrocarbon emissions in urban areas could exacerbate local air-pollution problems.

In sum, it is important to recognize that policies designed to prevent global warming and its adverse impacts may themselves have negative impacts. It is critical to design policies that effectively reduce greenhouse gas concentrations with a minimum of other adverse consequences.

CONCLUSIONS

The possibility of greenhouse warming presents California and the rest of the world with a plethora of problems. It also raises an overarching policy dilemma: while government should act to protect its citizens from the impacts of global warming, uncertainties inherent in the greenhouse issue make it difficult to justify instituting policies that themselves would have dramatic consequences on the economy and course of society.

There are three arguments to consider in dealing with this dilemma: First, many of the policies that will help prevent future warming can be justified for other reasons. For example, reduction of automobile, power-plant, and thermal-boiler emissions is already a main strategy for overcoming the staggering air-quality problems that plague California's urban areas. Further, measures that reduce petroleum use also reduce the state's vulnerability to the kind of shocks in oil prices and supplies experienced in the 1970s. Thus, increasing energy efficiency and switching to nonpolluting alternative power sources would bring immediate benefits to the state even without considering the issue of global warming (see table 1).

Second, Western society in general, and specifically California, has a tradition of insuring itself against potential events with drastic consequences. Even with a relatively small chance of disaster occurring, essentially every building of value in the state is insured for fire loss. The state's flood protection system has likewise cost billions of dollars. Yet the likelihood of a fire destroying any particular building during its useful life is very small; and there is only a fifty-fifty chance we will have a hundred-year flood during the next century. There seems to be a consensus among scientists that we face even odds of a substantial global warming during the next century, yet this threat has not elicited a response similar to those established for fire and flood. Part of the problem is that policymakers have seen the consequences of fires and floods but have not seen the consequences of a three-degree C warming. Yet if society is willing to trust its scientists in such potentially far-reaching matters as, say, genetic engineering or nuclear waste storage, should it not also recognize their warnings on global climate change?

Third, it is time for policymakers to start considering a broad range of issues related to the general topic of economic sustainability. Here sustainability refers to the natural resources (material inputs as well as the ability to absorb economic outputs like pollutants) that allow continuing any particular economic activity indefinitely into the future. Government presently seems to give little thought to the level of economic activity that can be sustained over the long term. But the threat of global warming dramatically illustrates that current trends in the use of global natural resources cannot continue indefinitely. And future warming is only one of many similar problems. It is clear, for example, that California's agriculture faces serious problems—for example, salinization and pesticide residues—that are independent of global warming but may equally threaten the state's agricultural productivity. California, along with the rest of the world, faces persistent and growing problems of pollution and resource depletion.

The environmental community and increasingly the business community agree that problems of environmental sustainability must be solved before they radically change the structure of our economy and society. Will we make a gradual transition to economic institutions that can be sustained, or will we continue with our current trends and wait for devastating changes to ravage the economy? Policymakers must recognize that resource and environmental issues will affect California's future in any event. They have an obligation to work toward a gradual rather than a radical transition. Does this mean that people must accept a lower standard of living? Not necessarily. Today's standard is now generally measured by the per capita value of all goods and services, but these measures do not necessarily recognize the value of improved efficiency; nor do they recognize the myriad nonmarket factors that affect people's lives. In short, we are using inadequate means to measure economic and social well-being.

With more realistic measures of economic well-being and with a strong sense of the importance of economic sustainability, policymakers can begin the arduous task of establishing new policies to deal with sustainability issues. The time is now.

RECOMMENDATIONS FOR STATE ACTION

- Conduct research to assess the full- and long-term social costs of greenhouse gas emissions.

- Expand research to evaluate the impact of future energy-cost changes on key economic activities within the state.
- Establish greenhouse gas emission reduction goals for California that will help achieve long-term energy and economic sustainability.
- Expand statewide programs to reduce energy demand and increase supply while still providing necessary energy-related services.
- Expand programs that encourage cost-effective water management, including increasing water-use efficiency and water-storage capabilities.
- Expand research to evaluate the opportunities for replacing fossil fuels with renewable biofuels.
- Expand research on the tree species and management techniques most appropriate for sequestering carbon in California. This research should emphasize forest management and reforestation systems as well as tree species that maximize carbon storage, optimize the production of usable fiber, and minimize atmospheric emissions of reactive hydrocarbons.
- Initiate research on the effects of state government policies on the economic and environmental sustainability of the state.

REFERENCES

Baxter, Lester W., and Kevin Calandri. 1989. The effects of global warming on electricity use in California. Paper presented at 9th Miami International Congress on Energy and the Environment, December 1989.

California Energy Commission. 1988. *Biennial conservation report.* October. Sacramento: California Energy Commission.

———. 1989a. *Biennial fuels report.* November. Sacramento: California Energy Commission.

———. 1989b. *California's energy agenda: The 1989 biennial report.* July. Sacramento: California Energy Commission.

Edwards, Allen G., et al. 1989. *The impacts of global warming on California.* Sacramento: California Energy Commission.

Energy Economist. July 1988.

Jaski, Mike, et al. 1989. *California energy demand 1989–2009.* December. Sacramento: California Energy Commission.

Smith, Joel B., and Dennis A. Tirpak, eds. 1988. *The potential effects of global climate change on the United States: Draft: Report to Congress.* October. Washington, D.C.: U.S. Environmental Protection Agency.

NINE

Studying the Human Dimensions of Global Climate Change

Richard A. Berk

SCIENTIFIC UNCERTAINTY AND THE GREENHOUSE EFFECT

Widespread concern about the environment is decades old. Within the past few years, however, dramatic events such as the discovery of the "ozone hole," advances in the natural sciences, and new research technology have significantly altered scientific perspectives on environmental issues. Nowhere is this more true than in the study of climate.

As the chapters in this volume suggest, with recent scientific interest in climate change has come a need to address substantive issues over very long periods of time and over virtually the entire globe. There is also a growing recognition not only of the links between physical and biological systems but also of the key roles played by human activities and institutions in interaction with the physical and biological world. Hence, the study of climate change presents a host of important questions to social scientists, for which they are not fully prepared.

In response, there have been over the past three years (1986–1989) a number of meetings and conferences in which research agendas for the "human dimensions" of global climate change have been addressed. Progress has been noticeable, but incremental. When, therefore, our July 1989 workshop on the "human dimensions" first convened, there was very little on which we could definitively build. The discussion was wide-ranging but somewhat diffuse, focused as much on reshaping the social sciences as on substantive topics. Both kinds of issues will be addressed below.

The problems inherent in studying the human dimensions of global climate change do not occur in a scientific vacuum. Rather, they are in part created by, and in part reflect, important gaps in scientific understanding of the physical and biological dimensions. To set the stage, therefore, the general nature of these gaps needs to be briefly reviewed.

The Physical and Biological Dimensions

There is little doubt that the greenhouse theory is correct in general terms.[1] However, it is sometimes not fully appreciated how many important details remain unclear.[2]

The role of clouds, for example, is very poorly understood: thus, projected temperature increases may vary from about one to five degrees from that source of uncertainty alone.[3] In addition, evidence that the increase in greenhouse gases is already producing demonstrable changes in climate is equivocal, and the rate at which temperatures will change over time is extremely difficult to forecast.[4] Among the many problems undermining credible future temperature forecasts are substantial scientific uncertainty about many important processes, data of uneven quality, and insufficient computing power.[5]

Furthermore, since temperature is the engine driving climate, prospective global warming may well have a number of secondary and tertiary physical effects and feedbacks. There may be, for instance, changes in the pattern and overall amount of precipitation, and for many physical processes, precipitation is at least as important as temperature.[6] For example, the polar ice sheets play an important role in climate projections, but a lot will depend upon how much snow falls in the polar regions and how much eventually melts. In other words, uncertainty about increases in temperature is hardly the only problem.

Within an overall pattern, moreover, there will no doubt be substantial variation across different microclimates.[7] Warmer temperatures in the American Southwest, for instance, may heat the deserts, but actually make areas along the Pacific Coast cooler and wetter because of deeper intrusion of the marine layer. Variation across microclimates is important not only because that is what living things experience but also because it is far from obvious how to go from the global models to microclimates—or even that the same sorts of formulations apply.

Finally, most of the scientific research, thus far, has focused on possible increases in mean temperature, yet much of the effect from global warming will be experienced as increases in *variability*, or in the number of *extreme events*.[8] For example, an increase of four degrees in average temperature may be far less important than a doubling from ten to twenty days of very high temperatures, placing stress on humans, animals, and plants ill adapted to cope with periodic extremes.[9] In other words, we must consider the role of changing *distributions* of various climate characteristics, temperature and precipitation distributions in particular. In this light, for many distributional forms (e.g., Gaussian), a relatively small increase in average temperature translates into a relatively large increase in the number of extreme events. That is, the current range of projected average temperature increases is for many distributional forms fully consistent with rather dramatic increases in the number of extreme events.

The general point is that despite excellent science and genuine advances, we are a long way from a detailed understanding of the physical and biological dimensions of global climate change. Particularly frustrating is that so much effort has been directed toward modeling central tendencies at a macro level, when for plants, animals, and humans what may well matter more is other distributional parameters at a micro level. In any case, the resulting uncertainties are multiplied when the human dimensions of global climate change are considered.

The Human Dimensions

The human dimensions of global climate change are an integral part of the scientific uncertainty surrounding the physical and biological dimensions of global climate change. On the one hand, human activity is obviously crucial for the production of greenhouse gases. The very things that help sustain human life—agriculture and manufacturing—are substantially responsible for the global warming that may well occur. On the other hand, global warming could affect a host of human activities in perhaps dramatic and life-threatening ways. Clearly, significant changes in temperature and precipitation would threaten current arrangements both within national boundaries and between nations. As a modest instance, imagine the consequences if the best regions in North America for growing wheat and corn were shifted approximately three hundred miles north.

The link between the impact of human activity on climate and the impact of climate on human activity is, in turn, a key to understanding, and perhaps influencing, public policy. Obviously, actual and/or projected climate effects are a critical stimulus for ameliorative policy instruments (whether for prevention or adaptation). But such stimuli are at best necessary conditions soon overlaid with social constructions that may transform the "objective" features of climate change into a variety of subjective meanings. Thus, for some, rapid global warming will be a painful payback for past abuses of the planet. For others, it will be a blessing fostering reforms that should have been undertaken in any case. For still others, it will be a windfall of economic and political opportunities.

In short, it is hard to imagine any area of human activity that is not bound up with climate and climate change. Indeed, the study of global climate change should significantly be a study of human activity. Unfortunately, the very pervasiveness of the relationships between climate and behavior is more than a rich opportunity: it is also a fundamental conceptual and practical obstacle.

To begin, it is important to stress that there is effectively no social science for the study of global climate change. By way of analogy, suppose that the small collection of natural science disciplines that currently study global change did not exist: climatology, meteorology, atmospheric sciences, and the like. In that situation, virtually all of the more traditional disciplines such as physics, chemistry, and biology would be the key resources and would presumably have something important to contribute. Consequently, to write an essay on what the traditional natural sciences had to offer for the study of global climate change, one would, in effect, need to discuss each discipline in considerable detail. In chemistry, for example, one would draw on several different subdisciplines: physical chemistry, organic chemistry, and biochemistry. Put another way, it is hard to imagine a kind of chemistry that would be unimportant. The same would hold for physics and biology.

Not that all of each discipline would be equally relevant; but within almost all of the traditional disciplinary divisions, there would be at least some relevant expertise. The trick would be to find it and organize it. Yet, since the importance of the expertise would actually vary, and since there might be some distance between the expertise and the climate problem being addressed, it would be very difficult to establish priorities. Indeed, if one knew enough to establish priorities, one would probably not have to undertake the exercise of establishing priorities to begin with. One would already be well under way.

The point is that much of the splendid science currently under way on climate owes its existence not just to the relative maturity of the physical and biological sciences but also to the rather recent development of hybrid scientific activities, which concentrate heterogeneous expertise on critical substantive problems and provide significant conceptual organization and infrastructure for the field. In the absence of that institutionalization, an essay on the physical or biological dimensions of global climate change would risk reading like a catalog of courses from the science departments of every major university in the country.

An essay on the human dimensions of global climate change runs the same risk. In the absence of a social science of global climate change, one might be tempted to ransack all of the heterogeneous social science disciplines and their interdisciplinary progeny seeking all possible contributions.[10] The result would be almost indistinguishable from a university course catalog:[11] Theory and Policy of Taxation; Industrial Organization, Price Policies, and Regulation; Economic Development and Culture Change; Group Behavior; Risk Analysis; Human Information Processing; Environmental Psychology; Public Opinion and Voter Behavior;

Political Psychology; Population and Fertility; Social Movements; and so on. Each of these specialties within the social sciences would in principle have something interesting to offer. But priorities would be very difficult to establish a priori and would more likely reflect the political influence of particular disciplines than any reasonable scientific criteria. On the issue of energy conservation, for example, how could one know in advance whether a greater investment should be made in applying utility theory from economics or human information processing from psychology? If one knew enough to decide, one would have already done the research.

These problems suggest that an essay on the human dimensions of global climate change would be more useful if focused on something other than the ubiquitous "review of the literature." There seem to be at least two productive options. First, it may be instructive to select a small number of substantive problems of obvious importance and potentially broad appeal. The use of land (e.g., deforestation/reforestation) is a popular illustration. This has been the strategy favored by a committee of the Social Science Research Council [12] and by a working group of the Committee on Global Change of the National Research Council (NRC).[13] In other words, some substantive research priorities are being suggested, but with the clear understanding that research problems not cited may still be of some interest; a lower priority is not to be equated with worthlessness. Second, it might be useful to suggest kinds of studies that should be pursued on the basis of research tradition or style, cutting across substantive problems and scholarly disciplines. Examples include social-indicators research, ethnographics, and program evaluation.

Both ways to establishing priorities have merit, and we will address each in turn. In so doing, we will build on the agenda-setting efforts of the Social Science Research Council, the National Research Council, and of course, on the discussions of the "Human Dimensions Panel," which met for three days during the workshop that inspired this volume.

SUBSTANTIVE PRIORITIES IN RESEARCH ON THE HUMAN DIMENSIONS OF GLOBAL CLIMATE CHANGE

Probably the most thorough and eclectic consideration of the substantive issues associated with the human dimensions of global climate change has been undertaken by the Social Science Research Council (SSRC). In several meetings held over a two-year period, the SSRC drew on the expertise of a large number of social, biological, and physical scientists to define a number of broad substantive problems of *interest to social scientists* as well as biological and physical scientists. This work led to a proposal, funded by the National Science Foundation, in which research is beginning on six substantive problems that overlap with those selected by the National Research Council's working group. In what follows, we draw heavily on the suggestions of the SSRC and NRC groups, although the priorities expressed reflect most directly the deliberations of the Davis workshop.

Land Use Changes [14]

While a number of topics might be productively addressed under the rubric of land use, the NRC's working group favored a set of research activities in three substantive areas, which are also consistent with the interests of the SSRC Committee. The overarching theme is that utilization of land is a central organizing factor in human societies and that how land is used has long-term environmental consequences.

1) Land cover trajectory scenarios for the next century on tropical forest deforestation/afforestation and wetland conversion, with an emphasis on wet rice cultivation. The purposes of these scenarios would be "1) to develop bounds and trajectories for future land transformations and their associated changes in physical and biological attributes and chemical emissions, and 2) to develop an understanding of the relative importance of different forces (population, institutions, technology, economic structure policies, environmental feedbacks, etc.) for land transformation." [15] In the medium term at least, the product would pose descriptions of alternative futures rather than sets of structural equations.

2) Global models of land transformations. The purpose of the global model would be to estimate the bounds of emissions and other changes in land-surface characteristics due to agricultural production over the next century, while responding to resource constraints (agro-ecological considerations) and socioeconomic factors. [16]

3) Process studies of biomass burning, livestock development, and urbanization, to improve basic understanding of the causes and magnitude of land use transformations. [17] For example, one might study large-scale composting practices in Scandinavia as much for what might be learned about kinds of incentives that seem effective as for what might be learned about the technology.

With some minor recasting, each of these three proposals could be enormously interesting and helpful for regional research within California. For example, urbanization continues throughout the state with unknown implications for land use and its effects on local micro-climates. An important place to begin working on formal regional modeling would be to determine what already exists in the way of regional macroeconomic models. In addition to models used for economic forecasting, there are models used to provide "social impact assessments" (often required by law for major land developments). Some of these models can be found in academe while others have been devised by private research firms, banks, and real estate developers. All such models are likely to be found wanting, but as much might be learned from their weaknesses as their strengths.

For example, any long-term projections of land use require reasonable approximations not just of population growth in general but of changing regional age pyramids. What matters is not just how many people but what kinds of people populate a given area: age and reproductive capacity are highly significant. However, age pyramid projections for regions within California require not just extrapolations from current birth figures but also plausible projections for in-migration and out-migration (for both the state as a whole and for particular regions), which are not easily produced.

Industrial Metabolism

Like land use, industrial "metabolism" covers a very rich menu of topics. As well, industrial metabolism is a central organizing factor in human society with long-term environmental consequences. Three kinds of studies in related areas were recommended by the NRC's working group.

1) Studies on the intensity of energy and materials use. Of particular importance would be research into changes over time and space in energy or material requirements (and their relative importance) to provide end-use services. The role of end-use services would be central in part because of the judgment that their role has not been given the attention it deserves.[18]

2) Research on the timing of industrial/technological change. Of particular importance would be spatial comparisons taking into account such things as economic growth, factor abundance, labor force productivity, absolute wealth, and cultural factors.[19]

3) Case studies on the evolution of industrial metabolism and its waste streams. Such studies would help provide better understanding of the shifts over time in the relative contributions of production versus consumption as sources of environmental perturbations.[20]

Again, translations for California are easily made. For example, we might hypothesize that, over the past several decades, the relative contribution of consumption versus production as a source for environmental stress has increased. Put another way, it can be argued that producers have been relatively more responsive to environmental threats (and governmental regulations) than consumers. In California, this issue could be addressed by case study research on major energy producers. For instance, what has been the percentage change over the past two decades in the production of greenhouse gases by companies engaged in oil extraction and refining compared with the production of greenhouse gases by private automobiles? Likewise, it would be informative to investigate how manufacturing firms compare with households in their recycling of energy, glass, or aluminum. Possibly the modification of today's consumer behavior could offer far greater potential for managing environmental stress than further expensive ultimatums to major producers.

Climate Change and Epidemiology

Diseases of various sorts are clearly linked to climate. The spread of disease is affected by complicated interactions among pathogens, vectors, hosts, susceptible organisms, and infected organisms,[21] with the numbers and resilience of each of these potentially influenced by climate. Lyme disease is one of the more interesting and challenging examples. For example, ticks capable of spreading Lyme disease can thrive in Humboldt County but struggle to survive in Kern County. Consequently, climate *change* may well affect the spread of disease. For example, rising water levels in rivers and streams caused by greater precipitation might detrimentally affect the breeding habitat of certain kinds of mosquitoes that transmit diseases such as viral meningitis. Alternatively, some kinds of mosquitoes actually breed more productively in moderately turbulent than in still water. In short, climate changes are likely to affect patterns of disease in plants, animals, and humans by altering the mix of pathogens and vectors.

In an important sense, however, these illustrations are too simple. Changing patterns of human activity—agricultural practices, residential arrangements, recreational activities, and others—affect the environment in which microorganisms and vectors live, and indirectly affect their viability.[22] For example, replacing a swamp with landfill will dramatically affect mosquito populations. Human activities also affect the kinds of risks to which people are exposed, the likelihood of person-to-person transmission, and the kinds of possible medical responses. Hiking and camping, for instance, place people at risk for Lyme disease. Generally speaking, many kinds of human activities will mediate, diffuse, or exacerbate the particular physical or biological impacts of climate change that, in turn, affect the spread of disease.

At least three important research activities might be undertaken under the heading of climate change and epidemiology.

1) Conceptualize the links between climate, human activities, and disease. Clearly, much more is known today about some pathogens than others; current knowledge should be organized so that needs for new information can begin to be met. Some of this work has already begun under the auspices of the Environmental Protection Agency.

2) Formalize epidemiological models with the potential to project both the likelihood and the course of future epidemics and to simulate the impact of various policy interventions. For example, suppose southern California became warmer and wetter, while immigration from Central America increased substantially. What would be the implications for malaria? Alternatively, what if all immigrants from Central America were screened for malaria and quarantined if infected, but returning tourists were not?[23]

3) Develop and rigorously evaluate public health measures selected in part by projecting consequences for California of climate change. For malaria, for instance, what combination of mosquito control and screening of immigrants and tourists would be most cost-effective?

Public Perceptions of Climate Change

The summer of 1988 showed us that policy responses to climate change will depend in part on a concerned and mobilized public. However, virtually no solid information exists about how local climate (or "microclimate") is *experienced*. Past research has focused on individuals' environmentally relevant values and on support for or opposition to various policy options, matters that are rather different from what concerns people day to day about climate. As noted previously, moreover, the focus of climate modelers on forecasts of central tendencies, rather than on forecasts of variances, extreme values, and strings of extreme values, has implied that people are concerned about the "typical" change. In fact, what may really concern them is the atypical or extreme events. Moreover, their responses may well be highly nonlinear or even discontinuous. Put somewhat too simply, a four-degree average increase in temperature over the course of a summer may go almost unnoticed. Even three days in a row with highs over 110 degrees may be annoying, but not a great source of discomfort or concern. However, ten consecutive days with highs over 110 may bring people to a breaking point. This phenomenon can be described rather more formally through a theory in psychophysics which suggests that response functions are often highly nonlinear.[24] The key point is really that human response toward different dimensions of climate is essentially unknown. Two kinds of research needs are implied.

1) Rigorous survey research to understand people's experience with climate. At the very least, research should examine people's reactions to different temperature and precipitation distributions, considering the four distributional properties we have noted: central tendency, variability, extreme events, and sequences of extreme events. In practice, this task might be accomplished through factorial survey methods[25] by which respondents are asked to react to "vignettes" or "scenarios" depicting a hypothetical climatic composite. For example, a given climate vignette might include a high mean temperature, low variance around the mean, and a relatively large number of very hot days scattered throughout the entire summer. However, the particular configuration of climate dimensions represented in the factorial surveys would be assigned at random, much as in a fractional factorial experiment.[26] Among the many possible responses that might be included along with the climate vignettes would be affective reactions (e.g., degree of discomfort), fears about climate in the future, belief in the role of greenhouse gases, and support for or opposition to different public policies.

2) **Laboratory research on how climate is experienced.** One might construct experimental settings in which temperature and humidity were under the experimenter's control and subject to wide and independent variation. A fractional factorial experimental design might involve the distributional parameters just discussed (central tendency, variability, extreme values, and strings of extreme values). An advantage of laboratory research over survey methods is that the stimulus to which the subject reacts is far more grounded. Nevertheless, exposure to variation in temperature and humidity in a laboratory is quite different from being exposed to such variations in the "real world." It would be interesting to see if the vignette stimulus and the laboratory stimulus produced the same general findings. For example, both might show that concerns about the greenhouse effect would be most strongly affected by strings of extremes and least strongly affected by central tendencies, with the impact of variation falling somewhere in between.

Behavior Change of Individuals and Organizations

With any acknowledgment that the greenhouse effect will be of practical significance comes a call for substantial changes in human activities. If the 1989 recommendations of the Los Angeles Air Quality Management District are a representative example, such changes will be difficult to implement. They include, for example, positive incentives to encourage retrofitting of inefficient home heating units, negative incentives to discourage unnecessary driving, and banning the sale of bias ply tires (which are less fuel-efficient than radial tires). Unfortunately, there is a very large gap between what is known about how to change human behavior in policy-constructive ways and the behavioral assumptions beneath the sorts of reform that are currently being linked to the greenhouse effect. Several high-priority research initiatives appear to be needed.

1) **Process studies of how household activities are organized.** For example, research on "household production" has, using the factory metaphor, stressed relationships between the input of labor and market goods with the output of economic well-being.[27] Hence, the "household production function" is treated by and large as a black box. What is needed is far more emphasis on what is inside the box, building in part on earlier time-use research.[28] A variety of methods would be appropriate, including observational studies, diary studies, and surveys. But unless there is better understanding about how people produce household commodities, it will be very difficult to design and implement household behavior-change instruments.

2) Process studies of how organizational activities are structured.
The arguments made for households apply to organizations. The emphasis here is less on the technology of production itself (which was discussed earlier) than on the ways in which the organization of the workday affects the production of greenhouse gases. For example, what are the options for particular kinds of businesses with regard to carpooling, van pooling, staggered work hours, or work at home? Just as in the case of households, observational studies, diary studies, or surveys would be useful.

3) Studies on how people respond to risk and incentives, combining recent work from cognitive psychology with work from economics. It is by now quite clear that much of conventional utility theory holds only approximately, and that sometimes the approximation is very poor.[29] Perhaps the groundwork has been laid for research on behavior-change strategies relevant to climate-change concerns which will be able to manipulate risk and incentives in ways more consistent with how people process data. For example, the same objective risk might be expressed in several different ways (e.g., as a probability versus a rate) to determine which elicits the most appropriate behavior.

4) Studies on how to shift demand away from high greenhouse gas activities. Decades of research in social psychology have demonstrated that, in principle, it is possible to change people's preferences. However, with few exceptions[30] these have not been field-tested in real conservation programs. For example, it seems that individual motivation to conserve must be based on the conviction that others drawing on the same resource pool will conserve as well. More broadly stated, conservation behavior has as much to do with "social pressure" as with assessments of risk or gaming solutions to common dilemmas. But how can such information be effectively conveyed in the real world of metered gas, water, and electricity? And if conveyed, would it really make a practical difference? Clearly, this is a ripe area for experimental research outside the laboratory, linked to real policy instruments.

METHODOLOGICAL PRIORITIES IN RESEARCH ON THE HUMAN DIMENSIONS OF GLOBAL CLIMATE CHANGE

In addition to setting substantive priorities, it may be useful to establish methodological priorities. That is, there may be certain styles of research that are especially effective in studies of the human dimensions of global climate change. These may also have implications for the appropriate research infrastructure.

Social-Indicators Data on Climate-Relevant Human Activity
It is patently obvious that a critical source of data on global warming is time-series measures on temperature and the concentrations of green-

house gases. That is, if something important is happening over time, it is vital to have over-time measures. The same logic holds for social phenomena. It is vital that data collection begin on a number of climate-relevant social indicators. Examples include use of mass transportation systems, the extent of ridesharing, when and how different kinds of home appliances are used, garden- and lawn-watering practices, the extent of paper recycling, composting, and the disposal of solid waste. In the context of "waste stream studies" mentioned above, for instance, it would be useful to plot time trends of the fraction of consumers who recycle glass bottles.[31]

Before such data may be properly collected, however, a great deal of developmental work would be required. What activities should be monitored and in what depth? What sort of geographical coverage is required (e.g., oversampling in environmentally sensitive locales)? What data collection technology should be employed (e.g., surveys, utility company records)? How often should measures be collected (e.g., yearly)? In short, there is the need not only for "social indicators" research on environmentally relevant activities but also for prior conceptual work.

What is required, therefore, is a planning group to design an indicators system. A number of important decisions with long-term consequences have to be made; researchers cannot redefine an indicator, its measures, or its database without effectively breaking the continuity in the time series. Thus consensus is required on key design issues which a wide variety of potential data users can live with for years to come (with users including not just researchers but political agencies, lobbying groups, and corporations). Ideally, the rough equivalent of "hearings" should be held to learn what sort of social/environmental indicators might be most useful.

Randomized Field Experiments in Behavior Change

Whether we take a preventive or adaptive position on the environment (or something in between), millions of Californians will have to change their behavior. While there is considerable theory in economics and social psychology about how to change people's environmentally relevant behavior, there have been very few large-scale social experiments in which efforts to change "real world" behavior (in contrast with laboratory studies) have been rigorously tried.[32] Perhaps the best-known experiments are the "peak load" pricing trials done in Los Angeles some time ago.[33] What seems needed, therefore, is a number of rigorous (with random assignment to experimental and control groups), large-scale social experiments tackling real behavior with real policy instruments. Examples of relevant activities include glass recycling, home furnace retrofitting, carpooling, and water conservation. Since, as a practical matter, social reforms tend to come in clusters, it would also be terribly important to employ factorial designs in which different *packages* of instruments were evaluated. The experimental units could be individuals or organizations such as schools, factories, or hospitals.

In contrast with the proposed program of climate-related social indicators, we are probably already able to launch randomized field trials with a number of pilot/feasibility studies. The conceptual and technical groundwork is more advanced,[34] and, unlike the collection of time-series data, changes can be made from experiment to experiment.

For example, financial incentives could be offered for a few months to a company's employees (perhaps employees at one of the UC campuses) to carpool or use existing van services. The pilot study would determine how best to deliver those incentives (e.g., cash versus amenities) and what level of incentives might be required. Also important would be to determine whether random assignment to different levels of incentives is ethical and practical. Finally, the pilot study would also require construction of initial measurement instruments, subject to later revision.

Modeling at the Regional, State, and Local Level

As noted earlier, a large number of modeling efforts already are in place throughout the region and state, which might serve as a foundation for new modeling projects on the relationships between microclimates and human activities. To the best of our knowledge, existing models are currently either forecasting devices, which are sometimes little more than fancy extrapolation procedures for current longitudinal databases, or more elaborate attempts to represent a number of causal mechanisms, typically in economic terms. Both modeling traditions are useful (and sometimes overlap) and can be built upon. In particular, model builders need 1) to introduce microclimate variables far more richly and explicitly, and 2) to include a number of noneconomic variables. The noneconomic variables might be measures of individual or organizational behavior (e.g., recycling), individual and organizational preferences (e.g., support for conservation on value grounds), and/or the structural arrangements within households (e.g., one wage earner or two) and in work organizations (e.g., provision of parking at work).

CONCLUSIONS:
IMPLICATIONS FOR RESEARCH INFRASTRUCTURE

Efforts at the national level are beginning to generate some support for research on the human dimensions of global climate change. In addition, agenda-setting initiatives by the Social Science Research Council and the National Research Council are facilitating contacts between social and natural scientists interested in the topic. However, given the absence of a social science of climate change (discussed earlier) and a history of small-scale, fragmented research projects, it should not be surprising that the research infrastructure in this area is woefully inadequate. Much needs to be done, and much could be done even at a local or regional level.

Data Archive

The human dimensions of global climate change necessarily touch on a wide variety of social phenomena: energy use, waste disposal, water supply, transportation, and many others. Some data of considerable relevance already exist, often in the administrative records of utility companies, government agencies, and private firms. For example, details on local water use can be retrieved from the billing records of water districts, while the purchase of energy-efficient appliances can be explored in sales records and perhaps market research from major manufacturers. In addition, the views of Californians on environmentally relevant matters have been studied fairly regularly by a number of local polling agencies. In short, there already exists a rich body of data that might be productively mined. However, no one apparently knows *exactly* what data resources for California already exist, what their general attributes are, and whether they are (or might be made) available for research purposes. Clearly, a survey of relevant data resources needs to be undertaken. Once the survey of data resources has been completed, the useful data would have to be obtained, thoroughly documented, and then stored in forms for easy retrieval and dissemination. This is no small task; it would require staff, computing facilities, and space, and there would have to be means for continually updating the archive as new data are collected.

There is a pressing need for this kind of research data archive on the human dimensions of global climate change, but focused on California and the western region of the United States and combined with relevant data on physical and biological climate processes. A case in point is the current controversy over energy use in the Los Angeles Basin. Policy analysts would benefit from specific information about the consumption of electricity, natural gas, oil, gasoline, and diesel fuel, but also from information about appliance purchases, carpooling, the use of air conditioning, and a host of other individual and organizational activities. Much of this database exists in various forms (if only in primary records) and could perhaps be organized in monthly averages amenable to longitudinal analysis. It could then be used in social-indicator research to depict energy-use trends over time and perhaps forecast into the future. The database could also be used to examine the impacts of various policy initiatives in the past, and would serve as one baseline for impact assessments of policy changes yet to be introduced.

An Interdisciplinary Research Center

A balanced research portfolio on the human dimensions of global climate change would significantly involve climate researchers of various kinds and contain a variety of social science research styles. With respect to the former, if political responses to the greenhouse effect depend heavily on what happens in the tails of the temperature/precipitation distribution, social scientists should confer with climate modelers on the parameters whose values they are forecasting (and for what geographical areas). Consultation might alter somewhat the models being used and/or the analysis of the output from these models. For example, a projected time series of average monthly temperatures could be analyzed with a number of different distributional parameters in mind.

With respect to research styles, there should be routine interactions between the formal modelers, the database archivists, the experimentalists, and the process researchers. In the past, these research traditions have gone their separate ways, to the disadvantage of all. In contrast, for instance, any experimental results from the field on the price elasticity for gasoline should be made available immediately to modelers. Equally important, modelers should help shape the particular questions that field experiments address: the key interventions to test (e.g., existing pricing versus possible increases in the gasoline tax) based on "high leverage points" in the model; the different "doses" of the intervention to be implemented (e.g., five cents per gallon, ten cents per gallon); the choice of experimental site (e.g., one rural area and one urban area); and so on. Such interactions would improve the quality of both models and field experiments.

However, such interactions are easily sidetracked. A way to enhance productive multidisciplinary research cutting across research styles would be to house a number of such researchers at one location. An interdisciplinary research center would be one vehicle, perhaps patterned after University of California Organized Research Units (ORUs). The organized research unit could be the site for the data archive and have resources to support sabbatical research, postdoctoral positions, and the usual sorts of necessary hardware. The goal would *not* be to concentrate in one location a large number of research activities that would otherwise proceed, but to bring together *only* those research activities that would especially benefit from ongoing and substantial *interdisciplinary* collaboration.

Training and Retooling

The number of social, biological, or physical scientists doing research on the human dimensions of global climate change is very small. While the reasons for this situation are beyond the scope of this essay, there are clearly serious shortages of trained personnel in the short and even medium term (since there are few instructors to train students) to grapple with the potential problems of climate change. A number of responses might emerge, which are familiar to academic administrators: undergraduate and graduate curriculum restructuring, graduate student fellowships, postdoctoral support, seed money for research, sabbatical year support for retooling, and so on. Clearly, a mix of approaches is desirable, depending on the particular goals and talents already available. To suggest a somewhat innovative step, however, universities might offer a new master's degree awarded along with a Ph.D. in a conventional social science discipline. The new degree would be given for substantial course work on global climate change and both a master's thesis and Ph.D. dissertation on the human dimensions of global climate change. A result could be, for instance, a Ph.D. in economics with a master's in the economics of global climate change. As an interim measure, at least, the "complementary master's" strategy could improve the pool of research talent, would require relatively little restructuring of existing curriculum, and would produce new Ph.D.'s who could still find jobs, if they wished, within the conventional disciplines and academic departments.[35]

To summarize, serious research on the human dimensions of global climate change will take a major resource investment. Yet if we are to move beyond armchair theory and back-of-the-envelope calculations, the research is essential. Resources by themselves are obviously not enough. The resources need to be invested in research and training that are truly interdisciplinary and that represent a productive mix of different research styles. This means building an organizational structure not dominated by the preferences of an elite few, but not immobilized by the diverse preferences of the many. A careful balance must be struck between leadership by elites and widespread participation.

NOTES

1. Schneider 1988, 1989; Houghton and Woodwell 1989.
2. Suomi 1975; Ramanathan 1988.
3. Ramanathan et al. 1989; Cess et al. 1989.
4. Epstein 1982; Wigley, Angell, and Jones 1985; Hansen and Lebedeff 1987; Schneider 1989.
5. Solow 1989.
6. Lean and Warrilow 1989.
7. Hansen et al. 1988.
8. Ibid.
9. Mearns, Katz, and Schneider 1988; Roberts 1989.
10. There are, of course, many interdisciplinary fields in the social sciences (e.g., social psychology, political economy) and many interdisciplinary applied fields as well (e.g., marketing, urban planning). The point is that there is really no well-established field analogous to climatology.
11. This partial list was taken from the UCLA catalog.
12. The Social Science Research Council's Committee for Research on Global Environmental Change is currently composed of Richard A. Berk, William C. Clark, Harold K.

Jacobson, Diana Liverman, William D. Nordhaus, John F. Richards, Thomas C. Schelling, Stephen H. Schneider, Billie Lee Turner, and Edith Brown Weiss (chair).

13. The working group on "Human Interactions with Global Climate Change" has been composed of William C. Clark, Robert Kates (chair), Thomas Lee, Vernon Ruttan, and Billie Lee Turner. A much larger group, however, was convened in the fall of 1989 to flesh out the working group's proposed research agenda.

14. In this and the next section on industrial metabolism, much will be drawn from the excellent literature summary written by Vicki Norberg-Bohm (1989).

15. Myers 1980; Arnold 1987; Palo 1987; Stavins and Jaffe 1988.

16. Grainger 1984; Kallio et al. 1987; Parks and Alig 1988.

17. Wolman and Fournier 1987; Richards and Tucker 1988.

18. Williams, Larson, and Ross 1987; Herman, Ardekani, and Ausubel 1989.

19. Lee and Nakicenovic 1988; Ausubel 1989.

20. Ayres and Rod 1986; Ayres et al. 1988.

21. Bailey 1975.

22. McNeil 1976; Burnet and White 1972.

23. Bailey 1982.

24. Spence 1990.

25. Rossi and Nock 1982.

26. Fleiss 1986.

27. Becker 1981.

28. Robinson 1977; Berk and Berk 1979; Juster and Stafford 1985.

29. Tversky and Kahneman 1974; Freudenberg 1988; Camerer and Kunreuther 1989.

30. Berk et al. 1981.

31. There is a long research tradition on social indicators in the social sciences upon which to build (Land and Spilerman 1975; Land and Juster 1981; MacRae 1985).

32. Over the past two decades, there have been hundreds of randomized field experiments on other substantive problems (Rossi and Freeman 1989).

33. Aigner 1985.

34. Hausman and Wise 1985; Fleiss 1986.

35. For more details, see Berk and Clark 1990.

REFERENCES

Aigner, Dennis J. 1985. The residential electricity time-of-use pricing experiments: What have we learned? In *Social experimentation*, ed. J. A. Hausman and D. A. Wise. Chicago: University of Chicago Press.

Arnold, J. E. M. 1987. Deforestation. In *Resources and world development*, ed. D. J. McLaren and B. J. Skinner. Chichester, England: John Wiley and Sons.

Ausubel, J. H. 1989. Regularities in technological development: An environmental view. In *Technology and environment*, ed. J. H. Ausubel and H. E. Sladovich. Washington, D.C.: National Academy Press.

Ayres, R. U., and S. R. Rod. 1986. Reconstructing an environmental history: Patterns of pollution in the Hudson-Raritan basin. *Environment* 28, no. 4:14–20, 39–43.

Ayres, R. U., L. W. Ayres, J. A. Tarr, and R. C. Widgery. 1988. An historical reconstruction of major pollutant levels in the Hudson-Raritan basin: 1880–1980. NOAA Technical Memorandum NOS OMA 42. Washington, D.C.: United States Department of Commerce, National Oceanic and Atmospheric Administration.

Bailey, Norman T. J. 1975. *The mathematical theory of infectious diseases.* 2d ed. London: Charles Griffin and Co.

———. 1982. *The biomathematics of malaria.* New York: Oxford University Press.

Becker, G. S. 1981. *A treatise on the family.* Cambridge, Mass.: Harvard University Press.

Berk, Richard A., and Sarah Fenstermaker Berk. 1979. *Labor and leisure at home: Content and organization of the household day.* Beverly Hills: Sage Publications.

———. 1983. The supply-side sociology of the family: The challenge of the new home economics. *Annual Review of Sociology* 9:375–395.

Berk, R. A., and W. A. V. Clark. 1990. Teaching for the future: An essay on training social scientists for the study of global environmental change. Manuscript prepared for the Center for the Study of the Environment and Society, University of California, Los Angeles.

Berk, Richard A., C. J. LaCivita, Katherine Sredl, and Thomas F. Cooley. 1981. *Water shortage: Lesson in conservation from the great California drought, 1976–1977.* Cambridge, Mass.: Abt Books.

Burnet, F. M., and D. O. White. 1972. *The natural history of infectious disease.* 4th ed. Cambridge, Mass.: Cambridge University Press.

Camerer, Colin F., and Howard Kunreuther. 1989. Decision processes for low probability events: Policy implications. *Journal of Policy Analysis and Management* 8, no. 4:565–592.

Cess, R. D., G. L. Potter, J. P. Blanchet, G. J. Boer, S. J. Ghan, J. T. Kiehl, H. Le Treut, Z.-X. Li, X.-Z. Liang, J. F. B. Mitchell, J. J. Morcrette, D. A. Randall, M. R. Riches, E. Roeckner, U. Schlese, A. Slingo, K. E. Taylor, W. M. Washington, R. T. Wetherald, and I. Yagai. 1989. Interpretation of cloudclimate feedback as produced by 14 atmospheric general circulation models. *Science* 245:513–516.

Epstein, E. S. 1982. Detecting climate change. *Applied Meteorology* 21:1172–1182.

Fleiss, J. L. 1986. *The design and analysis of clinical experiments.* New York: John Wiley and Sons.

Freudenberg, William R. 1988. Perceived risk, real risk: Social science and the art of probabilistic risk assessment. *Science* 242:44–49.

Grainger, Alan. 1984. Quantifying changes in forest cover in the humid tropics: Overcoming current limitations. *Journal of World Forest Resource Management*, vol. 1, no. 1.

Hansen, J., and S. Lebedeff. 1987. Global trends of measured surface air temperature. *Journal of Geophysical Research*, vol. 92, no. D11:13,345–13,372.

Hansen, J., I. Fung, A. Lacis, D. Rind, S. Lebedeff, R. Ruedy, G. Russell, and P. Stone. 1988. Global climate changes as forecast by Goddard Institute for Space Studies three-dimensional model. *Journal of Geophysical Research*, vol. 93, no. D8:9341–9364.

Hausman, Jerry A., and David A. Wise, eds. 1985. *Social experimentation.* Chicago: University of Chicago Press.

Herman, R., S. A. Ardekani, and J. H. Ausubel. 1989. Dematerialization. In *Technology and environment*, ed. J. H. Ausubel and H. E. Sladovich. Washington, D.C.: National Academy Press.

Houghton, Richard A., and George M. Woodwell. 1989. Global climatic change. *Scientific American* 260, no. 4:36–44.

Juster, F. Thomas, and Frank P. Stafford, eds. 1985. *Time, goods, and well-being.* Ann Arbor: Survey Research Center, Institute for Social Research, University of Michigan.

Kallio, M., D. P. Dykstra, and C. S. Binkley, eds. 1987. *Global forest sector.* Chichester, England: John Wiley and Sons.

Land, K. C., and F. T. Juster. 1981. Social accounting systems: An overview. In *Social accounting systems: Essays on the state of the art,* ed. K. C. Land and F. T. Juster. New York: Academic Press.

Land, K. C., and S. Spilerman. 1975. *Social indicator models.* New York: Russell Sage Foundation.

Lean, J., and D. A. Warrilow. 1989. Simulation of regional climate impact on Amazon deforestation. *Nature* 342:411–413.

Lee, T. H., and N. Nakicenovic. 1988. Technology life-cycles and business decisions. *International Journal of Technology Management,* vol. 3, no. 4.

MacRae, Jr., Duncan. 1985. *Policy indicators.* Chapel Hill: University of North Carolina Press.

McNeil, William H. 1976. *Plagues and people.* New York: Anchor Press.

Mearns, L. O., R. W. Katz, and S. H. Schneider. 1984. Extreme high-temperature events: Changes in their probabilities with changes in mean temperature. *Journal of Climatology and Applied Meteorology* 23:1601–1613.

Myers, Norman. 1980. *Conversion of tropical moist forests.* Washington, D.C.: National Academy of Sciences.

Norberg-Bohm, V. 1989. A selective literature review on the human sources of global environmental change. In *Research Strategies for the U.S. Global Change Research Program,* U.S. National Research Council Committee on Global Change. Washington, D.C.: National Academy Press.

Palo, M. 1987. Deforestation perspectives for the tropics: A provisional theory with pilot applications. In *Global forest sector,* ed. M. Kallio, D. P. Dykstra, and C. S. Binkley. Chichester, England: John Wiley and Sons.

Parks, P. J., and R. J. Alig. 1988. Land base models for forest resource supply analysis: A critical review. *Canadian Journal of Forest Research,* vol. 18.

Ramanathan, V. 1988. The greenhouse theory of climate change: A test by an inadvertent global experiment. *Science* 240:293–299.

Ramanathan, V., R. D. Cess, E. F. Harrison, P. Minnis, B. R. Barkstrom, E. Ahmad, and D. Hartmann. 1989. Cloud radiative forcing and climate: Results from the earth radiation budget experiment. *Science* 243:57–63.

Richards, J. R., and Richard P. Tucker. 1988. *World deforestation in the twentieth century.* Durham, N.C.: Duke University Press.

Roberts, Leslie. 1989. Global warming: Blaming the sun. *Science* 246:992–993.

Robinson, J. P. 1977. *How Americans use time.* New York: Praeger.

Rossi, P. H., and H. E. Freeman. 1989. *Evaluation: A systematic approach.* 4th ed. Newbury Park, Calif.: Sage Publications.

Rossi, P. H., and S. L. Nock, eds. 1982. *Measuring social judgments.* Beverly Hills: Sage Publications.

Schneider, Stephen H. 1988. Climate modeling. *Scientific American* 256, no. 5: 72–80.

———. 1989. The greenhouse effect: Science and policy. *Science* 243:771–781.

Solow, Andrew R. 1989. Is it getting stuffy in here, or is it just my imagination? *Change* 2, no. 3:40–46.

Spence, I. 1990. The visual psychophysics of graphical elements. *Journal of Experimental Psychology: Human Perception and Performance* 16, no. 4:683–692.

Stavins, R. N., and Adam B. Jaffe. 1988. *Forested wetland depletion in the United States: An analysis of unintended consequences of federal policy and programs.* Discussion Paper Number 1391. Cambridge, Mass.: Harvard Institute of Economic Research.

Suomi, Verner E. 1975. *Understanding climatic change: A program for action.* Washington, D.C.: National Academy of Sciences.

Tversky, Amos, and Daniel Kahneman. 1974. Judgment under uncertainty: Heuristics and biases. *Science* 185:1124–1131.

Wigley, T. M. L., James K. Angell, and P. D. Jones. 1985. Analysis of the temperature record. In *Detecting the climatic effects of increasing carbon dioxide,* ed. Michael MacCracken and Frederick M. Luther. ER-0235. Washington, D.C.: U.S. Department of Energy.

Wildavsky, Aaron. 1989. *Searching for safety.* New Brunswick: Transaction Publishers.

Williams, R. H., Eric D. Larson, and Marc H. Ross. 1987. Materials, affluence, and industrial energy use. In *Annual Review of Energy,* 1987. Palo Alto, Calif.: Annual Reviews Press.

Wolman, M. G., and F. G. A. Fournier, eds. 1987. *Land transformation in agriculture.* SCOPE 32. New York: John Wiley and Sons.

TEN

Epilogue

Joseph B. Knox and Noreen G. Dowling

In retrospect, the year 1989 was a milestone year in the communities concerned with global change and climate research. Because extraordinary weather in 1988 brought enhanced awareness of the so-called "greenhouse effect" to the public, 1989 saw the initiation of a growing and dynamic Global Change Research Program within the U.S. government, increased fiscal support for coordinated research, and the pronouncement by President George Bush that the 1990s would be the decade of the environment. A cascade of "global warming" conferences and workshops appeared on the national and international scene and attracted public attention. Active media participation publicized new research results as well as ever stronger divergent opinions about the seriousness of the threat. A universal call went out for top-quality focused science and development of national policy based on sound research.

The three UC/DOE workshops conducted from July to October 1989 were only a small part of this spectrum of effort. Several significant trends related to global change converged during those same months. These included:

1) an emerging awareness by scientists that risk assessment methodologies and communication resources should be focused on the issues of global change;

2) a call by the director of DOE Office of Energy Research for quality research definitively linking greenhouse emissions with climate change; and

3) a call for a much stronger component of academic research on global environmental change to undergird the DOE Global Change Research Program and the tasks assigned to it.

In response to these converging thrusts, the U.S. House of Representatives Subcommittee on Energy and Water, on September 29, 1989, established the rationale and the funding for a National Institute for Global Environmental Change (NIGEC). Perhaps largely because of the Davis workshops, this legislation designated the University of California, Davis, as the management entity for a new national institute with regional research centers in four parts of the country—Northeast, Midwest, South, and West. Congressional interest and the ensuing events of organization eventually developed these regional centers at Harvard University, the University of Indiana, Tulane University, and the University of California. Thus was NIGEC conceptualized and born.

On October 19, 1989, NIGEC management and institutional representatives met with DOE representatives to explore research priorities and design an initial series of start-up research projects. NIGEC management and UC representatives in the Western Regional Center proposed a list of projects gleaned from the research agenda developed at the first UC/DOE workshop, "Global Climate Change and Its Effects on California," from which the papers in this volume appear. These papers had provided not only a springboard for discussion and meaningful exchange but also a practical list from which pertinent and seminal projects for the UC system could be selected. Further, many of UC's most qualified researchers, with their premier facilities and experience, had been marshaled through the workshop process to focus closely on the research needs of the sponsoring agency (DOE), and the UC systemwide workshop advisory committee had served as a resource and review body in the development of the project list. Once these initial projects are under way, the NIGEC Western Center will reach out to other regional institutions to request their participation in a yet broader research program.

We look forward with enthusiasm to the evolution and growth of NIGEC. Its work will contribute to the strengthened scientific understanding of global climate and environmental change and to the emergence of reasonable policy responses to the challenge our world faces. In addition, because one of the needs of the 1990s—the decade of the environment—will be the acute necessity for training in number and quality the scientists and engineers who must confront an uncertain future, many young researchers will be encouraged and trained through project work performed under NIGEC auspices. Together with many others we join in creatively confronting the challenges of the coming century—and building toward a more sustainable future.

CONTRIBUTORS

Barbara Bentley	Visiting Professor, Bodega Marine Laboratory, University of California. Currently Professor, Department of Ecology and Evolution, SUNY, Stony Brook, New York
Richard A. Berk	Professor, Department of Sociology, University of California, Los Angeles
Susan Bicknell	Professor of Forest Ecology, Department of Forestry, Humboldt State University
Daniel B. Botkin	Professor, Department of Biological Sciences, University of California, Santa Barbara
Noreen G. Dowling	Director, Public Service Research and Dissemination Program, and Associate Director, National Institute for Global Environmental Change, University of California, Davis
Allen G. Edwards	Senior Economist, California Energy Commission, State of California
Wayne Ferren	Curator, Herbarium, Department of Biological Sciences, University of California, Santa Barbara
W. Lawrence Gates	Chief Scientist, Program for Climate Model Diagnosis and Intercomparison, Lawrence Livermore National Laboratory, Livermore, California
Lynne Kennedy	Program Analyst, Agricultural Issues Center, University of California, Davis
Joseph B. Knox	Director, National Institute for Global Environmental Change, University of California, Davis

Lowell Lewis	Associate Vice President—Programs, Division of Agriculture and Natural Resources, Office of the President, University of California
Michael C. MacCracken	Division Leader, Atmospheric and Geophysical Sciences Division, Lawrence Livermore National Laboratory, Livermore, California
Robert A. Nisbet	Research Associate, Environmental Studies Program, University of California, Santa Barbara
William Rains	Professor, Department of Agronomy and Range Science, University of California, Davis
F. Sherwood Rowland	Professor, Department of Chemistry, University of California, Irvine, and Co-Director, Western Regional Center of the National Institute for Global Environmental Change, University of California
Henry J. Vaux, Jr.	Director, Water Resources Center, and Professor, Department of Resource Economics, University of California, Riverside
Charles Woodhouse	Curator, Santa Barbara Museum of Natural History, Santa Barbara, California

AUSPICES

The contributions to this volume were in part supported by the Workshop Process funded by the U.S. Department of Energy and executed by the University of California with leadership from the Public Service Research and Dissemination Program (UC Davis). The editing of the book manuscript was supported largely by the U.S. Department of Energy through the National Institute for Global Environmental Change.

Designer: U. C. Press Staff
Compositor: G & S Typesetters, Inc.
Text: 10/12 Baskerville
Display: Baskerville
Printer: Maple-Vail
Binder: Maple-Vail